A Handbook of Statistical Analyses using SAS

SECOND EDITION

A Handbook of Statistical Analyses using SAS

SECOND EDITION

Geoff Der
Statistician
MRC Social and Public Health Sciences Unit
University of Glasgow
Glasgow, Scotland

and

Brian S. Everitt
Professor of Statistics in Behavioural Science
Institute of Psychiatry
University of London
London, U.K.

CHAPMAN & HALL/CRC

Boca Raton London New York Washington, D.C.

Library of Congress Cataloging-in-Publication Data

Der, Geoff.
A handbook of statistical analyses using SAS / G. Der and B.S. Everitt.—2nd ed.
p. cm.
Includes bibliographical references and index.
ISBN 1-58488-245-X (alk. paper)
1. SAS (Computer file) 2. Mathematical statistics—Data processing. I. Everitt, Brian. II. Title.
QA276.4 .D47 2001
519.5'0285—dc21 2001037656

Direct all inquiries to CRC Press LLC, 2000 N.W. Corporate Blvd., Boca Raton, Florida 33431.

Visit the CRC Press Web site at www.crcpress.com

International Standard Book Number 1-58488-245-X
Library of Congress Card Number 2001037656
Printed in the United States of America 5 6 7 8 9 0
Printed on acid-free paper

Preface

SAS, standing for Statistical Analysis System, is a powerful software package for the manipulation and statistical analysis of data. The system is extensively documented in a series of manuals. In the first edition of this book we estimated that the relevant manuals ran to some 10,000 pages, but one reviewer described this as a considerable underestimate. Despite the quality of the manuals, their very bulk can be intimidating for potential users, especially those relatively new to SAS. For readers of this edition, there is some good news: the entire documentation for SAS has been condensed into one slim volume — a Web browseable CD-ROM. The bad news, of course, is that you need a reasonable degree of acquaintance with SAS before this becomes very useful.

Here our aim has been to give a brief and straightforward description of how to conduct a range of statistical analyses using the latest version of SAS, version 8.1. We hope the book will provide students and researchers with a self-contained means of using SAS to analyse their data, and that it will also serve as a "stepping stone" to using the printed manuals and online documentation.

Many of the data sets used in the text are taken from *A Handbook of Small Data Sets* (referred to in the text as *SDS*) by Hand et al., also published by Chapman and Hall/CRC.

The examples and datasets are available on line at: http://www.sas.com/service/library/onlinedoc/code.samples.html.

We are extremely grateful to Ms. Harriet Meteyard for her usual excellent word processing and overall support during the preparation and writing of this book.

Geoff Der
Brian S. Everitt

Contents

Chapter 1

A Brief Introduction to SAS

1.1 Introduction

The SAS system is an integrated set of modules for manipulating, analysing, and presenting data. There is a large range of modules that can be added to the basic system, known as BASE SAS. Here we concentrate on the STAT and GRAPH modules in addition to the main features of the base SAS system.

At the heart of SAS is a programming language composed of statements that specify how data are to be processed and analysed. The statements correspond to operations to be performed on the data or instructions about the analysis. A SAS program consists of a sequence of SAS statements grouped together into blocks, referred to as "steps." These fall into two types: data steps and procedure (proc) steps. A data step is used to prepare data for analysis. It creates a SAS data set and may reorganise the data and modify it in the process. A proc step is used to perform a particular type of analysis, or statistical test, on the data in a SAS data set.

A typical program might comprise a data step to read in some raw data followed by a series of proc steps analysing that data. If, in the course of the analysis, the data need to be modified, a second data step would be used to do this.

The SAS system is available for a wide range of different computers and operating systems and the way in which SAS programs are entered and run differs somewhat according to the computing environment. We describe the Microsoft Windows interface, as this is by far the most popular, although other windowing environments, such as X-windows, are quite similar.

1.2 The Microsoft Windows User Interface

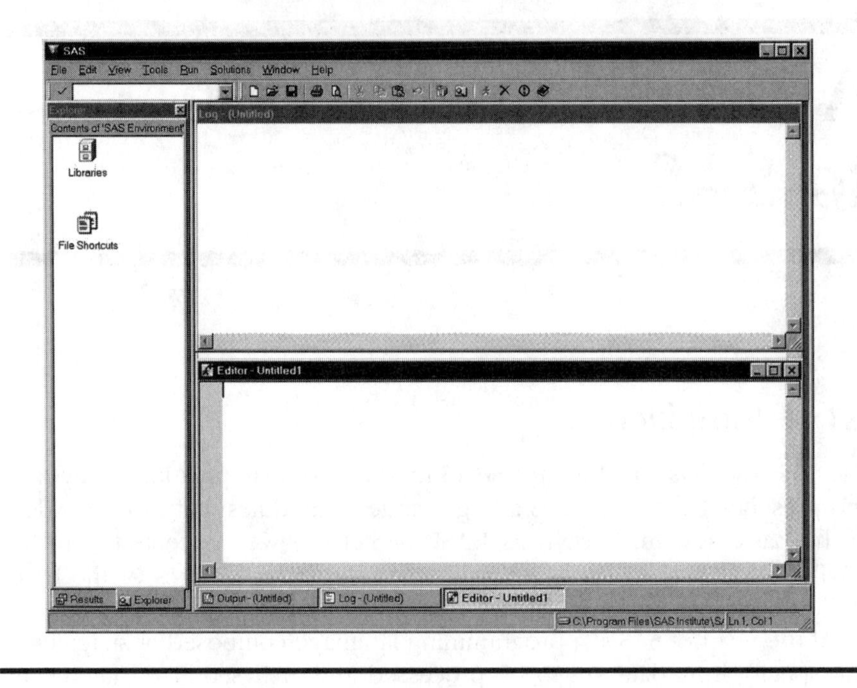

Display 1.1

Display 1.1 shows how SAS version 8 appears running under Windows.

When SAS is started, there are five main windows open, namely the Editor, Log, Output, Results, and Explorer windows. In Display 1.1, the Editor, Log, and Explorer windows are visible. The Results window is hidden behind the Explorer window and the Output window is hidden behind the Program Editor and Log windows.

At the top, below the SAS title bar, is the menu bar. On the line below that is the tool bar with the command bar at its left end. The tool bar consists of buttons that perform frequently used commands. The command bar allows one to type in less frequently used commands. At the bottom, the status line comprises a message area with the current directory and editor cursor position at the right. Double-clicking on the current directory allows it to be changed.

Briefly, the purpose of the main windows is as follows.

1. Editor: The Editor window is for typing in editing, and running programs. When a SAS program is run, two types of output are generated: the log and the procedure output, and these are displayed in the Log and Output windows.
2. Log: The Log window shows the SAS statements that have been submitted together with information about the execution of the program, including warning and error messages.
3. Output: The Output window shows the printed results of any procedures. It is here that the results of any statistical analyses are shown.
4. Results: The Results window is effectively a graphical index to the Output window useful for navigating around large amounts of procedure output. Right-clicking on a procedure, or section of output, allows that portion of the output to be viewed, printed, deleted, or saved to file.
5. Explorer: The Explorer window allows the contents of SAS data sets and libraries to be examined interactively, by double-clicking on them.

When graphical procedures are run, an additional window is opened to display the resulting graphs.

Managing the windows (e.g., moving between windows, resizing them, and rearranging them) can be done with the normal windows controls, including the **Window** menu. There is also a row of buttons and tabs at the bottom of the screen that can be used to select a window. If a window has been closed, it can be reopened using the **View** menu.

To simplify the process of learning to use the SAS interface, we concentrate on the Editor, Log, and Output windows and the most important and useful menu options, and recommend closing the Explorer and Results windows because these are not essential.

1.2.1 The Editor Window

In version 8 of SAS, a new editor was introduced, referred to as the enhanced editor. The older version, known as the program editor, has been retained but is not recommended. Here we describe the enhanced editor and may refer to it simply as "the editor." If SAS starts up using the program editor rather than the enhanced editor, then from the **Tools** menu select **Options**; **Preferences** then the **Edit** tab and select the **Use Enhanced Editor** option*.

* At the time of writing, the enhanced editor was not yet available under X-windows.

The editor is essentially a built-in text editor specifically tailored to the SAS language and with additional facilities for running SAS programs.

Some aspects of the Editor window will be familiar as standard features of Windows applications. The File menu allows programs to be read from a file, saved to a file, or printed. The File menu also contains the command to exit from SAS. The Edit menu contains the usual options for cutting, copying, and pasting text and those for finding and replacing text.

The program currently in the Editor window can be run by choosing the Submit option from the Run menu. The Run menu is specific to the Editor window and will not be available if another window is the active window. Submitting a program may remove it from the Editor window. If so, it can be retrieved by choosing Recall Last Submit from the Run menu.

It is possible to run part of the program in the Editor window by selecting the text and then choosing Submit from the Run menu. With this method, the submitted text is not cleared from the Editor window. When running parts of programs in this way, make sure that a full step has been submitted. The easiest way to do this is to include a Run statement as the last statement.

The Options submenu within Tools allows the editor to be configured. When the Enhanced Editor window is the active window (View, Enhanced Editor will ensure that it is), Tools; Options; Enhanced Editor Options will open a window similar to that in Display 1.2. The display shows the recommended setup, in particular, that the options for collapsible code sections and automatic indentation are selected, and that Clear text on submit is not.

1.2.2 The Log and Output Windows

The contents of the Log and Output windows cannot be edited; thus, several options of the File and Edit menus are disabled when these windows are active.

The Clear all option in the Edit menu will empty either of these windows. This is useful for obtaining a "clean" printout if a program has been run several times as errors were being corrected.

1.2.3 Other Menus

The View menu is useful for reopening a window that has been closed.

The Solutions menu allows access to built-in SAS applications but these are beyond the scope of this text.

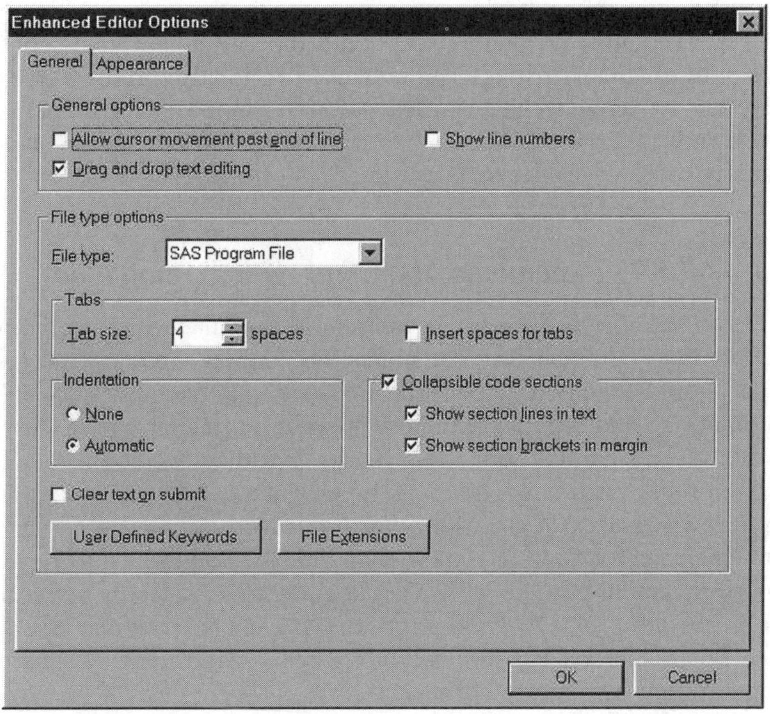

Display 1.2

The **Help** menu tends to become more useful as experience in SAS is gained, although there may be access to some tutorial materials if they have been licensed from SAS. Version 8 of SAS comes with a complete set of documentation on a CD-ROM in a format that can be browsed and searched with an HTML (Web) browser. If this has been installed, it can be accessed through **Help**; **Books and Training**; **SAS Online Doc**.

Context-sensitive help can be invoked with the F1 key. Within the editor, when the cursor is positioned over the name of a SAS procedure, the F1 key brings up the help for that procedure.

1.3 The SAS Language

Learning to use the SAS language is largely a question of learning the statements that are needed to do the analysis required and of knowing how to structure them into steps. There are a few general principles that are useful to know.

Most SAS statements begin with a keyword that identifies the type of statement. (The most important exception is the assignment statement that begins with a variable name.) The enhanced editor recognises keywords as they are typed and changes their colour to blue. If a word remains red, this indicates a problem. The word may have been mistyped or is invalid for some other reason.

1.3.1 All SAS Statements Must End with a Semicolon

The most common mistake for new users is to omit the semicolon and the effect is to combine two statements into one. Sometimes, the result will be a valid statement, albeit one that has unintended results. If the result is not a valid statement, there will be an error message in the SAS log when the program is submitted. However, it may not be obvious that a semicolon has been omitted before the program is run, as the combined statement will usually begin with a valid keyword.

Statements can extend over more than one line and there may be more than one statement per line. However, keeping to one statement per line, as far as possible, helps to avoid errors and to identify those that do occur.

SAS statements fall into four broad categories according to where in a program they can be used. These are

1. Data step statements
2. Proc step statements
3. Statements that can be used in both data and proc steps
4. Global statements that apply to all following steps

Because the functions of the data and proc steps are so different, it is perhaps not surprising that many statements are only applicable to one type of step.

1.3.2 Program Steps

Data and proc steps begin with a **data** or **proc** statement, respectively, and end at the next **data** or **proc** statement, or the next **run** statement. When a data step has the data included within it, the step ends after the data. Understanding where steps begin and end is important because SAS programs are executed in whole steps. If an incomplete step is submitted, it will not be executed. The statements that were submitted will be listed in the log, but SAS will appear to have stopped at that point without explanation. In fact, SAS will simply be waiting for the step to be completed before running it. For this reason it is good practice to explicitly mark the end of each step by inserting a **run** statement and especially important to include one as the last statement in the program.

The enhanced editor offers several visual indicators of the beginning and end of steps. The **data**, **proc**, and **run** keywords are colour-coded in Navy blue, rather than the standard blue used for other keywords. If the enhanced editor options for collapsible code sections have been selected as shown in Display 1.2, each data and proc step will be separated by lines in the text and indicated by brackets in the margin. This gives the appearance of enclosing each data and proc step in its own box.

Data step statements must be within the relevant data step, that is, after the **data** statement and before the end of the step. Likewise, proc step statements must be within the proc step.

Global statements can be placed anywhere. If they are placed within a step, they will apply to that step and all subsequent steps until reset. A simple example of a global statement is the **title** statement, which defines a title for procedure output and graphs. The title is then used until changed or reset.

1.3.3 Variable Names and Data Set Names

In writing a SAS program, names must be given to variables and data sets. These can contain letters, numbers, and underline characters, and can be up to 32 characters in length but cannot begin with a number. (Prior to version 7 of SAS, the maximum length was eight characters.) Variable names can be in upper or lower case, or a mixture, but changes in case are ignored. Thus **Height**, **height**, and **HEIGHT** would all refer to the same variable.

1.3.4 Variable Lists

When a list of variable names is needed in a SAS program, an abbreviated form can often be used. A variable list of the form **sex - - weight** refers to the variables **sex** and **weight** and all the variables positioned between them in the data set. A second form of variable list can be used where a set of variables have names of the form **score1, score2, ... score10**. That is, there are ten variables with the root **score** in common and ending in the digits 1 to 10. In this case, they can be referred to by the variable list **score1 - score10** and do *not* need to be contiguous in the data set.

Before looking at the SAS language in more detail, the short example shown in Display 1.3 can be used to illustrate some of the preceding material. The data are adapted from Table 17 of *A Handbook of Small Data Sets* (*SDS*) and show the age and percentage body fat for 14 women. Display 1.4 shows how the example appears in the Editor window. The Results and Explorer windows have been closed and the Editor window maximized. The program consists of three steps: a data step followed by two proc steps. Submitting this program results in the log and procedure output shown in Displays 1.5 and 1.6, respectively.

From the log one can see that the program has been split into steps and each step run separately. Notes on how the step ran follow the statements that comprise the step. Although notes are for information only, it is important to check them. For example, it is worth checking that the notes for a data step report the expected number of observations and variables. The log may also contain warning messages, which should always be checked, as well as error messages.

The reason the log refers to the SAS data set as WORK.BODYFAT rather than simply bodyfat is explained later.

```
data bodyfat;
   Input age pctfat;
datalines;
23  28
39  31
41  26
49  25
50  31
53  35
53  42
54  29
56  33
57  30
58  33
58  34
60  41
61  34
;
proc print data=bodyfat;
run;
proc corr data=bodyfat;
run;
```

Display 1.3

SAS - [ch1ex1.sas]

File Edit View Tools Run Solutions Window Help

```
data bodyfat;
   Input age pctfat;
datalines;
23  28
39  31
41  26
49  25
50  31
53  35
53  42
54  29
56  33
57  30
58  33
58  34
60  41
61  34
;
proc print data=bodyfat;
run;
proc corr data=bodyfat;
run;
```

Output - (Untitled) | Log - (Untitled) | ch1ex1.sas

C:\Program Files\SAS Institute\S Ln 1, Col 1

Display 1.4

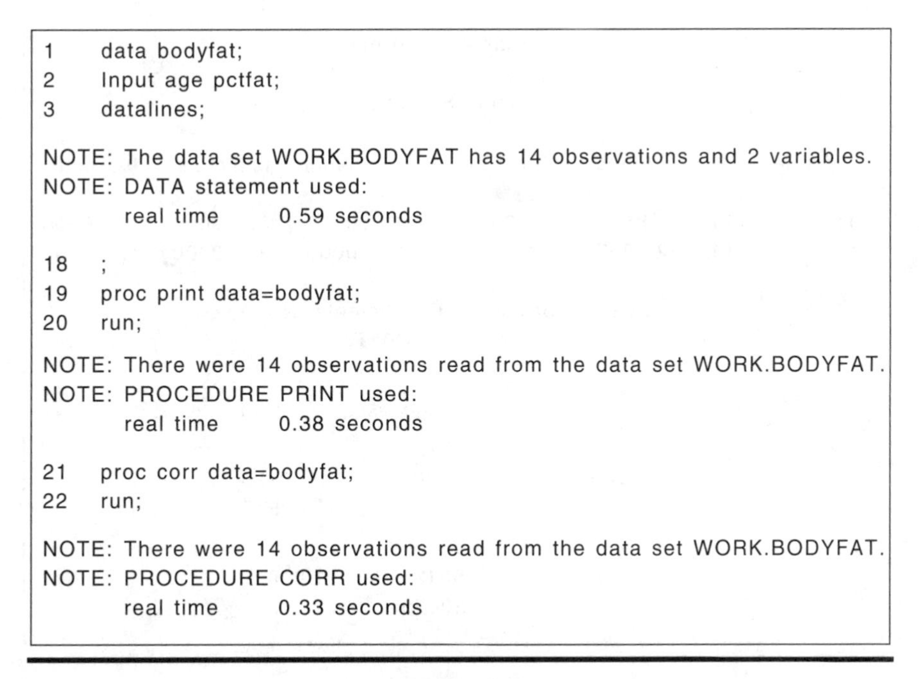

```
1      data bodyfat;
2      Input age pctfat;
3      datalines;

NOTE: The data set WORK.BODYFAT has 14 observations and 2 variables.
NOTE: DATA statement used:
         real time        0.59 seconds

18   ;
19   proc print data=bodyfat;
20   run;

NOTE: There were 14 observations read from the data set WORK.BODYFAT.
NOTE: PROCEDURE PRINT used:
         real time        0.38 seconds

21   proc corr data=bodyfat;
22   run;

NOTE: There were 14 observations read from the data set WORK.BODYFAT.
NOTE: PROCEDURE CORR used:
         real time        0.33 seconds
```

Display 1.5

```
                      The SAS System                       1

                  Obs       age       pctfat

                    1        23         28
                    2        39         31
                    3        41         26
                    4        49         25
                    5        50         31
                    6        53         35
                    7        53         42
                    8        54         29
                    9        56         33
                   10        57         30
                   11        58         33
                   12        58         34
                   13        60         41
                   14        61         34

                      The SAS System                       2

                    The CORR Procedure

                 2 Variables: age pctfat

                    Simple Statistics

Variable   N       Mean   Std Dev          Sum   Minimum    Maximum

age       14   50.85714  10.33930    712.00000  23.00000   61.00000
pctfat    14   32.28571   4.92136    452.00000  25.00000   42.00000

           Pearson Correlation Coefficients, N = 14
                  Prob > |r| under H0: Rho=0

                             age          pctfat

             age         1.00000         0.50125
                                          0.0679

             pctfat      0.50125         1.00000
                          0.0679
```

Display 1.6

1.4 The Data Step

Before data can be analysed in SAS, they need to be read into a SAS data set. Creating a SAS data set for subsequent analysis is the primary function of the data step. The data can be "raw" data or come from a previously created SAS data set. A data step is also used to manipulate, or reorganise the data. This can range from relatively simple operations (e.g., transforming variables) to more complex restructuring of the data. In many practical situations, organising and preprocessing the data takes up a large portion of the overall time and effort. The power and flexibility of SAS for such data manipulation is one of its great strengths.

We begin by describing how to create SAS data sets from raw data and store them on disk before turning to data manipulation. Each of the subsequent chapters includes the data step used to prepare the data for analysis and several of them illustrate features not described in this chapter.

*1.4.1 Creating SAS Data Sets from Raw Data**

Display 1.7 shows some hypothetical data on members of a slimming club, giving the membership number, team, starting weight, and current weight. Assuming these are in the file wgtclub1.dat, the following data step could be used to create a SAS data set.

```
data wghtclub;
   infile 'n:\handbook2\datasets\wgtclub1.dat';
   input idno team $ startweight weightnow;
run;
```

```
1023   red     189   165
1049   yellow  145   124
1219   red     210   192
1246   yellow  194   177
1078   red     127   118
1221   yellow  220   .
1095   blue    135   127
1157   green   155   141
```

* A "raw" data file can also be referred to as a text file, or ASCII file. Such files only include the printable characters plus tabs, spaces, and end-of-line characters. The files produced by database programs, spreadsheets, and word processors are not normally "raw" data, although such programs usually have the ability to "export" their data to such a file.

```
1331   blue     187   172
1067   green    135   122
1251   blue     181   166
1333   green    141   129
1192   yellow   152   139
1352   green    156   137
1262   blue     196   180
1087   red      148   135
1124   green    156   142
1197   red      138   125
1133   blue     180   167
1036   green    135   123
1057   yellow   146   132
1328   red      155   142
1243   blue     134   122
1177   red      141   130
1259   green    189   172
1017   blue     138   127
1099   yellow   148   132
1329   yellow   188   174
```

Display 1.7

1.4.2 The Data Statement

The data statement often takes this simple form where it merely names the data set being created, in this case wghtclub.

1.4.3 The Infile Statement

The infile statement specifies the file where the raw data are stored. The full pathname of the file is given. If the file is in the current directory (i.e., the one specified at the bottom right of the SAS window), the file name could have been specified simply as 'wghtclub1.dat'. Although many of the examples in this book use the shorter form, the full pathname is recommended. The name of the raw data file must be in quotes. In many cases, the infile statement will only need to specify the filename, as in this example.

In some circumstances, additional options on the **infile** statement will be needed. One such instance is where the values in the raw data file are not separated by spaces. Common alternatives are files in which the data values are separated by tabs or commas. Most of the raw data files used in later chapters are taken directly from the *Handbook of Small Data Sets*, where the data values are separated by tabs. Consequently, the **expandtabs** option, which changes tab characters into a number of spaces, has been used more often than would normally be the case. The **delimiter** option can be used to specify a separator. For example, **delimiter=','** could be used for files in which the data values are separated by commas. More than one delimiter can be specified. Chapter 6 contains an example.

Another situation where additional options may be needed is to specify what happens when the program requests more data values than a line in the raw data file contains. This can happen for a number of reasons, particularly where character data are being read. Often, the solution is to use the **pad** option, which adds spaces to the end of each data line as it is read.

1.4.4 The Input Statement

The **input** statement in the example specifies that four variables are to be read in from the raw data file: **idno**, **team**, **startweight**, and **weightnow**, and the dollar sign ($) after **team** indicates that it is a character variable. SAS has only two types of variables: numeric and character.

The function of the **input** statement is to name the variables, specify their type as numeric or character, and indicate where in the raw data the corresponding data values are. Where the data values are separated by spaces, as they are here, a simple form of the **input** statement is possible in which the variable names are merely listed in order and character variables are indicated by a dollar sign ($) after their name. This is the so-called "list" form of input. SAS has three main modes of input:

- List
- Column
- Formatted

(There is a fourth form — named input — but data suitable for this form of input occur so rarely that its description can safely be omitted.)

List input is the simplest and is usually to be preferred for that reason. The requirement that the data values be separated by spaces has some important implications. The first is that missing values cannot be represented by spaces in the raw data; a period (.) should be used instead. In the example, the value of **weightnow** is missing for member number 1221.

The second is that character values cannot contain spaces. With list input, it is also important to bear in mind that the default length for character variables is 8.

When using list input, always examine the SAS log. Check that the correct number of variables and observations have been read in. The message: "SAS went to a new line when INPUT statement reached past the end of a line" often indicates problems in reading the data. If so, the pad option on the infile statement may be needed.

With small data sets, it is advisable to print them out with proc print and check that the raw data have been read in correctly.

If list input is not appropriate, column input may be. Display 1.8 shows the slimming club data with members' names instead of their membership numbers.

To read in the data in the column form of input statement would be

```
input name $ 1-18 team $ 20-25 startweight 27-29 weightnow
31-33;
```

```
David Shaw          red     189  165
Amelia Serrano      yellow  145  124
Alan Nance          red     210  192
Ravi Sinha          yellow  194  177
Ashley McKnight     red     127  118
Jim Brown           yellow  220
Susan Stewart       blue    135  127
Rose Collins        green   155  141
Jason Schock        blue    187  172
Kanoko Nagasaka     green   135  122
Richard Rose        blue    181  166
Li-Hwa Lee          green   141  129
Charlene Armstrong  yellow  152  139
Bette Long          green   156  137
Yao Chen            blue    196  180
Kim Blackburn       red     148  135
Adrienne Fink       green   156  142
Lynne Overby        red     138  125
John VanMeter       blue    180  167
Becky Redding       green   135  123
Margie Vanhoy       yellow  146  132
Hisashi Ito         red     155  142
Deanna Hicks        blue    134  122
```

```
Holly Choate        red     141   130
Raoul Sanchez       green   189   172
Jennifer Brooks     blue    138   127
Asha Garg           yellow  148   132
Larry Goss          yellow  188   174
```

Display 1.8

As can be seen, the difference between the two forms of **input** statement is simply that the columns containing the data values for each variable are specified after the variable name, or after the dollar in the case of a character variable. The start and finish columns are separated by a hyphen; but for single column variables it is only necessary to give the one column number.

With formatted input, each variable is followed by its input format, referred to as its informat. Alternatively, a list of variables in parentheses is followed by a format list also in parentheses. Formatted input is the most flexible, partly because a wide range of informats is available. To read the above data using formatted input, the following **input** statement could be used:

```
input name $19. team $7. startweight 4. weightnow 3.;
```

The informat for a character variable consists of a dollar, the number of columns occupied by the data values and a period. The simplest form of informat for numeric data is simply the number of columns occupied by the data and a period. Note that the spaces separating the data values have been taken into account in the informat.

Where numeric data contain an implied decimal point, the informat has a second number after the period to indicate the number of digits to the right of the decimal point. For example, an informat of 5.2 would read five columns of numeric data and, in effect, move the decimal point two places to the left. Where the data contain an explicit decimal point, this takes precedence over the informat.

Formatted input must be used if the data are not in a standard numeric format. Such data are rare in practice. The most common use of special SAS informats is likely to be the date informats. When a date is read using a date informat, the resultant value is the number of days from January 1st 1960 to that date. The following data step illustrates the use of the **ddmmyyw.** informat. The width **w** may be from 6 to 32 columns. There is also the **mmddyyw.** informat for dates in American format. (In addition, there are corresponding output formats, referred to simply as "formats" to output dates in calendar form.)

```
data days;
input day ddmmyy8.;
cards;
231090
23/10/90
23 10 90
23101990
;
run;
```

Formatted input can be much more concise than column input, particularly when consecutive data values have the same format. If the first 20 columns of the data line contain the single-digit responses to 20 questions, the data could be read as follows:

```
input (q1 - q20) (20*1.);
```

In this case, using a numbered variable list makes the statement even more concise. The informats in the format list can be repeated by prefixing them with n*, where n is the number of times the format is to be repeated (20 in this case). If the format list has fewer informats than there are variables in the variable list, the entire format list is reused. Thus, the above input statement could be rewritten as:

```
input (q1 - q20) (1.);
```

This feature is useful where the data contain repeating groups. If the answers to the 20 questions occupied one and two columns alternately, they could be read with:

```
input (q1 - q20) (1. 2.);
```

The different forms of input can be mixed on the same input statement for maximum flexibility.

Where the data for an observation occupies several lines, the slash character (/), used as part of the input statement, indicates where to start reading data from the next line. Alternatively, a separate input statement could be written for each line of data because SAS automatically goes on to the next line of data at the completion of each input statement. In some circumstances, it is useful to be able to prevent SAS from automatically going on to the next line and this is done by adding an @ character to the end of the **input** statement. These features of data input are illustrated in later chapters.

1.4.5 Reading Data from an Existing SAS Data Set

To read data from a SAS data set, rather than from a raw data file, the **set** statement is used in place of the **infile** and **input** statements. The statement

```
data wgtclub2;
   set wghtclub;
run;
```

creates a new SAS data set **wgtclub2** reading in the data from **wghtclub**. It is also possible for the new data set to have the same name; for example, if the **data** statement above were replaced with

```
data wghtclub;
```

This would normally be used in a data step that also modified the data in some way.

1.4.6 Storing SAS Data Sets on Disk

Thus far, all the examples have shown temporary SAS data sets. They are temporary in the sense that they will be deleted when SAS is exited. To store SAS data sets permanently on disk and to access such data sets, the **libname** statement is used and the SAS data set referred to slightly differently.

```
libname db 'n:\handbook2\sasdata';
data db.wghtclub;
   set wghtclub;
run;
```

The **libname** statement specifies that the *libref* **db** refers to the directory **'n:\handbook2\sasdata'**. Thereafter, a SAS data set name prefixed with **'db.'** refers to a data set stored in that directory. When used on a **data** statement, the effect is to create a SAS data set in that directory. The data step reads data from the temporary SAS data set **wghtclub** and stores it in a permanent data set of the same name.

Because the libname statement is a global statement, the link between the *libref* db and the directory n:\handbook2\sasdata remains throughout the SAS session, or until reset. If SAS has been exited and restarted, the libname statement will need to be submitted again.

In Display 1.5 we saw that the temporary data set bodyfat was referred to in the log notes as 'WORK.BODYFAT'. This is because work is the *libref* pointing to the directory where temporary SAS data sets are stored. Because SAS handles this automatically, it need not concern the user.

1.5 Modifying SAS Data

As well as creating a SAS data set, the data step can also be used to modify the data in a variety of ways.

1.5.1 Creating and Modifying Variables

The assignment statement can be used both to create new variables and modify existing ones. The statement

```
weightloss=startweight-weightnow;
```

creates a new variable weigtloss and sets its value to the starting weight minus the current weight, and

```
startweight=startweight * 0.4536;
```

will convert the starting weight from pounds to kilograms.

SAS has the normal set of arithmetic operators: +, -, / (divide), * (multiply), and ** (exponentiate), plus various arithmetic, mathematical, and statistical functions, some of which are illustrated in later chapters.

The result of an arithmetic operation performed on a missing value is itself a missing value. When this occurs, a warning message is printed in the log. Missing values for numeric variables are represented by a period (.) and a variable can be set to a missing value by an assignment statement such as:

```
age = . ;
```

To assign a value to a character variable, the text string must be enclosed in quotes; for example:

```
team='green';
```

A missing value can be assigned to a character variable as follows:

```
Team='';
```

To modify the value of a variable for some observations and not others, or to make different modifications for different groups of observations, the **assignment** statement can be used within an **if then** statement.

```
reward=0;
if weightloss > 10 then reward=1;
```

If the condition **weigtloss > 10** is true, then the **assignment** statement **reward=1** is executed; otherwise, the variable **reward** keeps its previously assigned value of 0. In cases like this, an **else** statement could be used in conjunction with the **if then** statement.

```
if weightloss > 10 then reward=1;
   else reward=0;
```

The condition in the **if then** statement can be a simple comparison of two values. The form of comparison can be one of the following:

Operator		*Meaning*	*Example*
EQ	=	Equal to	a = b
NE	~=	Not equal to	a ne b
LT	<	Less than	a < b
GT	>	Greater than	a gt b
GE	>=	Greater than or equal to	a >= b
LE	<=	Less than or equal to	a le b

Comparisons can be combined into a more complex condition using **and** (&), **or** (|), and **not**.

```
if team='blue' and weightloss gt 10 then reward=1;
```

In more complex cases, it may be advisable to make the logic explicit by grouping conditions together with parentheses.

Some conditions involving a single variable can be simplified. For example, the following two statements are equivalent:

```
if age > 18 and age < 40 then agegroup = 1;
if 18 < age < 40 then agegroup = 1;
```

and conditions of the form:

```
x = 1 or x = 3 or x = 5
```

can be abbreviated to

```
x in(1, 3, 5)
```

using the in operator.

If the data contain missing values, it is important to allow for this when recoding. In numeric comparisons, missing values are treated as smaller than any number. For example,

```
if age >= 18 then adult=1;
   else adult=0;
```

would assign the value 0 to adult if age was missing, whereas it may be more appropriate to assign a missing value. The missing function could be used do this, by following the else statement with:

```
if missing(age) then adult=.;
```

Care needs to be exercised when making comparisons involving character variables because these are case sensitive and sensitive to leading blanks.

A group of statements can be executed conditionally by placing them between a do statement and an end statement:

```
If weightloss > 10 and weightnow < 140 then do;
target=1;
reward=1;
team ='blue';
end;
```

Every observation that satisfies the condition will have the values of target, reward, and team set as indicated. Otherwise, they will remain at their previous values.

Where the same operation is to be carried out on several variables, it is often convenient to use an array and an iterative do loop in combination. This is best illustrated with a simple example. Suppose we have 20 variables, q1 to q20, for which "not applicable" has been coded -1 and we wish to set those to missing values; we might do it as follows:

```
array qall {20} q1-q20;
do i= 1 to 20;
   if qall{i}=-1 then qall{i}=.;
end;
```

The **array** statement defines an array by specifying the name of the array, **qall** here, the number of variables to be included in braces, and the list of variables to be included. All the variables in the array must be of the same type, that is, all numeric or all character.

The iterative do loop repeats the statements between the **do** and the **end** a fixed number of times, with an index variable changing at each repetition. When used to process each of the variables in an array, the do loop should start with the index variable equal to 1 and end when it equals the number of variables in the array.

The array is a shorthand way of referring to a group of variables. In effect, it provides aliases for them so that each variable can be referred to by using the name of the array and its position within the array in braces. For example, **q12** could be referred to as **qall{12}** or when the variable **i** has the value **12** as **qall{i}**. However, the array only lasts for the duration of the data step in which it is defined.

1.5.2 Deleting Variables

Variables can be removed from the data set being created by using the **drop** or **keep** statements. The **drop** statement names a list of variables that are to be excluded from the data set, and the **keep** statement does the converse, that is, it names a list of variables that are to be the only ones retained in the data set, all others being excluded. So the statement **drop x y z;** in a data step results in a data set that does not contain the variables **x**, **y**, and **z**, whereas **keep x y z;** results in a data set that contains only those three variables.

1.5.3 Deleting Observations

It may be necessary to delete observations from the data set, either because they contain errors or because the analysis is to be carried out on a subset of the data. Deleting erroneous observations is best done using the **if then** statement with the **delete** statement.

```
if weightloss > startweight then delete;
```

In a case like this, it would also be useful to write out a message giving more information about the observation that contains the error.

```
if weightloss > startweight then do;
put 'Error in weight data' idno= startweight= weightloss=;
delete;
end;
```

The put statement writes text (in quotes) and the values of variables to the log.

1.5.4 Subsetting Data Sets

If analysis of a subset of the data is needed, it is often convenient to create a new data set containing only the relevant observations. This can be achieved using either the subsetting if statement or the where statement. The subsetting if statement consists simply of the keyword if followed by a logical condition. Only observations for which the condition is true are included in the data set being created.

```
data men;
   set survey;
   if sex='M';
run;
```

The statement where sex='M'; has the same form and could be used to achieve the same effect. The difference between the subsetting if statement and the where statement will not concern most users, except that the where statement can also be used with proc steps, as discussed below. More complex conditions can be specified in either statement in the same way as for an if then statement.

1.5.5 Concatenating and Merging Data Sets

Two or more data sets can be combined into one by specifying them in a single set statement.

```
data survey;
   set men women;
run;
```

This is also a simple way of adding new observations to an existing data set. First read the data for the new cases into a SAS data set and then combine this with the existing data set as follows.

```
data survey;
   set survey newcases;
run;
```

1.5.6 Merging Data Sets: Adding Variables

Data for a study can arise from more than one source, or at different times, and need to be combined. For example, demographic details from a questionnaire may need to be combined with the results of laboratory tests. To deal with this situation, the data are read into separate SAS data sets and then combined using a merge with a unique subject identifier as a key. Assuming the data have been read into two data sets, **demographics** and **labtests**, and that both data sets contain the subject identifier **idnumber**, they can be combined as follows:

```
proc sort data=demographics;
   by idnumber;
proc sort data=labtests;
   by idnumber;
data combined;
   merge demographics (in=indem) labtest (in=inlab);
   by idnumber;
   if indem and inlab;
run;
```

First, both data sets must be sorted by the matching variable **idnumber**. This variable should be of the same type, numeric or character, and same length in both data sets. The **merge** statement in the data step specifies the data sets to be merged. The option in parentheses after the name creates a temporary variable that indicates whether that data set provided an observation for the merged data set. The **by** statement specifies the matching variable. The subsetting **if** statement specifies that only observations having both the demographic data and the lab results should be included in the combined data set. Without this, the combined data set may contain incomplete observations, that is, those where there are demographic data but no lab results, or vice versa. An alternative would be to print messages in the log in such instances as follows.

```
If not indem then put idnumber ' no demographics';
If not inlab then put idnumber ' no lab results';
```

This method of match merging is not confined to situations in which there is a one-to-one correspondence between the observations in the data sets; it can be used for one-to-many or many-to-one relationships as well. A common practical application is in the use of look-up tables. For example, the research data set might contain the respondent's postal code (or zip code), and another file contain information on the characteristics of the area. Match merging the two data sets by postal code would attach area information to the individual observations. A subsetting **if** statement would be used so that only observations from the research data are retained.

1.5.7 The Operation of the Data Step

In addition to learning the statements that can be used in a data step, it is useful to understand how the data step operates.

The statements that comprise the data step form a sequence according to the order in which they occur. The sequence begins with the **data** statement and finishes at the end of the data step and is executed repeatedly until the source of data runs out. Starting from the **data** statement, a typical data step will read in some data with an **input** or **set** statement and use that data to construct an observation. The observation will then be used to execute the statements that follow. The data in the observation can be modified or added to in the process. At the end of the data step, the observation will be written to the data set being created. The sequence will begin again from the **data** statement, reading the data for the next observation, processing it, and writing it to the output data set. This continues until all the data have been read in and processed. The data step will then finish and the execution of the program will pass on to the next step.

In effect, then, the data step consists of a loop of instructions executed repeatedly until all the data is processed. The automatic SAS variable, _n_, records the iteration number but is not stored in the data set. Its use is illustrated in later chapters.

The point at which SAS adds an observation to the data set can be controlled using the **output** statement. When a data step includes one or more **output** statements an observation is added to the data set each time an **output** statement is executed, but not at the end of the data step. In this way, the data being read in can be used to construct several observations. This is illustrated in later chapters.

1.6 The proc Step

Once data have been read into a SAS data set, SAS procedures can be used to analyse the data. Roughly speaking, each SAS procedure performs a specific type of analysis. The proc step is a block of statements that specify the data set to be analysed, the procedure to be used, and any further details of the analysis. The step begins with a proc statement and ends with a run statement or when the next data or proc step starts. We recommend including a run statement for every proc step.

1.6.1 The proc Statement

The proc statement names the procedure to be used and may also specify options for the analysis. The most important option is the data= option, which names the data set to be analysed. If the option is omitted, the procedure uses the most recently created data set. Although this is usually what is intended, it is safer to explicitly specify the data set.

Many of the statements that follow particular proc statements are specific to individual procedures and are described in later chapters as they arise. A few, however, are more general and apply to a number of procedures.

1.6.2 The var Statement

The var statement specifies the variables that are to be processed by the proc step. For example:

```
proc print data=wghtclub;
   var name team weightloss;
run;
```

restricts the printout to the three variables mentioned, whereas the default would be to print all variables.

1.6.3 The where Statement

The where statement selects the observations to be processed. The keyword where is followed by a logical condition and only those observations for which the condition is true are included in the analysis.

```
proc print data=wghtclub;
   where weightloss > 0;
run;
```

1.6.4 The by Statement

The by statement is used to process the data in groups. The observations are grouped according to the values of the variable named in the by statement and a separate analysis is conducted for each group. To do this, the data set must first be sorted in the by variable.

```
proc sort data=wghtclub;
   by team;
proc means;
   var weightloss;
   by team;
run;
```

1.6.5 The class Statement

The class statement is used with many procedures to name variables that are to be used as classification variables, or factors. The variables named can be character or numeric variables and will typically contain a relatively small range of discreet values. There may be additional options on the class statement, depending on the procedure.

1.7 Global Statements

Global statements can occur at any point in a SAS program and remain in effect until reset.

The title statement is a global statement and provides a title that will appear on each page of printed output *and* each graph until reset. An example would be:

```
title 'Analysis of Slimming Club Data';
```

The text of the title must be enclosed in quotes. Multiple lines of titles can be specified with the title2 statement for the second line, title3 for the third line, and so on up to ten. The title statement is synonymous with title1. Titles are reset by a statement of the form:

```
title2;
```

This will reset line two of the titles and all lower lines, that is, title3, etc.; and title1; would reset all titles.

Comment statements are global statements in the sense that they can occur anywhere. There are two forms of **comment** statement. The first form begins with an asterisk and ends with a semicolon, for example:

```
* this is a comment;
```

The second form begins with /* and ends with */.

```
/* this is also a
   comment
*/
```

Comments can appear on the same line as a SAS statement; for example:

```
bmi=weight/height**2;              /* Body Mass Index */
```

The enhanced editor colour codes comment green, so it is easier to see if the */ has been omitted from the end or if the semicolon has been omitted in the first form of comment.

The first form of comment is useful for "commenting out" individual statements, whereas the second is useful for commenting out one or more steps because it can include semicolons.

The **options** and **goptions** global statements are used to set SAS system options and graphics options, respectively. Most of the system options can be safely left at their default values. Some of those controlling the procedure output that can be considered useful include:

- **nocenter** Aligns the output at the left, rather than centering it on the page; useful when the output linesize is wider than the screen.
- **nodate** Suppresses printing of the date and time on the output.
- **ps=*n*** Sets the output pagesize to *n* lines long.
- **ls=*n*** Sets the output linesize to *n* characters.
- **pageno=*n*** Sets the page number for the next page of output (e.g., **pageno=1** at the beginning of a program that is to be run repeatedly).

Several options can be set on a single **options** statement; for example:

```
options nodate nocenter pagegno=1;
```

The **goptions** statement is analogous, but sets graphical options. Some useful options are described below.

1.8 ODS: The Output Delivery System

The Output Delivery System (ODS) is the facility within SAS for formatting and saving procedure output. It is a relatively complex subject and could safely be ignored (and hence this section skipped!). This book does not deal with the use of ODS to format procedure output, except to mention that it enables output to be saved directly in HTML, pdf, or rtf files*.

One useful feature of ODS is the ability to save procedure output as SAS data sets. Prior to ODS, SAS procedures had a limited ability to save output — parameter estimates, fitted values, residuals, etc. — in SAS data sets, using the **out=** option on the **proc** statement, or the **output** statement. ODS extends this ability to the full range of procedure output. Each procedure's output is broken down into a set of tables and one of these can be saved to a SAS data set by including a statement of the form

```
ods output table = dataset;
```

within the proc step that generates the output.

Information on the tables created by each procedure is given in the "Details" section of the procedure's documentation. To find the variable names, use **proc contents data=*dataset*;** or **proc print** if the data set is small. A simple example is given in Chapter 5.

1.9 SAS Graphics

If the SAS/GRAPH module has been licensed, several of the statistical procedures can produce high-resolution graphics. Where the procedure does not have graphical capabilities built in, or different types of graphs are required, the general-purpose graphical procedures within SAS/GRAPH may be used. The most important of these is the **gplot** procedure.

1.9.1 Proc gplot

The simplest use of **proc gplot** is to produce a scatterplot of two variables, **x** and **y** for example.

```
proc gplot;
   plot y * x;
run;
```

* Pdf and rtf files from version 8.1 of SAS onwards.

A wide range of variations on this basic form of plot can be produced by varying the **plot** statement and using one or more **symbol** statements. The default plotting symbol is a plus sign. If no other plotting symbol has been explicitly defined, the default is used and the result is a scatterplot with the data points marked by pluses. The **symbol** statement can be used to alter the plot character, and also to control other aspects of the plot. To produce a line plot rather than a scatterplot:

```
symbol1 i=join;
proc gplot;
   plot y * x;
run;
```

Here, the **symbol1** statement explicitly defines the plotting symbol and the **i** (**interpolation**) option specifies that the points are to be joined. The points will be plotted in the order in which they occur in the data set, so it is usually necessary to sort the data by the *x*-axis variable first.

The data points will also be marked with pluses. The **v=** (**value=**) option in the **symbol** statement can be used to vary or remove the plot character. To change the above example so that only the line is plotted without the individual points being marked, the **symbol** statement would be:

```
symbol1 v=none i=join;
```

Other useful variations on the plot character are: x, star, square, diamond, triangle, hash, dot, and circle.

A variation of the **plot** statement uses a third variable to plot separate subgroups of the data. Thus,

```
symbol1 v=square i=join;
symbol2 v=triangle i=join;
proc gplot;
plot y * x = sex;
run;
```

will produce two lines with different plot characters. An alternative would be to remove the plot characters and use different types of line for the two subgroups. The **l=** (**linetype**) option of the **symbol** statement may be used to achieve this; for example,

```
symbol1 v=none i=join l=1;
symbol2 v=none i=join l=2;
```

```
proc gplot;
plot y * x = sex;
run;
```

Both of the above examples assume that two symbol definitions are being generated — one by the symbol1 statement and the other by symbol2. However, this is not the case when SAS is generating colour graphics. The reason is that SAS will use the symbol definition on the symbol1 statement once for each colour currently defined before going on to use symbol2. If the final output is to be in black and white, then the simplest solution is to begin the program with:

```
goptions colors=(black);
```

If the output is to be in colour, then it is simplest to use the c= (color=) option on the symbol statements themselves. For example:

```
symbol1 v=none i=join c=blue;
symbol2 v=none i=join c=red;
proc gplot;
plot y * x = sex;
run;
```

An alternative is to use the repeat (r=) option on the symbol statement with r=1. This is also used for the opposite situation, to force a symbol definition to be used repeatedly.

To plot means and standard deviations or standard errors, the i=std option can be used. This is explained with an example in Chapter 10.

Symbol statements are global statements and thus remain in effect until reset. Moreover, all the options set in a symbol statement persist until reset. If a program contains the statement

```
symbol1 i=join v=diamond c=blue;
```

and a later symbol statement

```
symbol1 i=join;
```

the later plot will also have the diamond plot character as well as the line, and they will be coloured blue.

To reset a symbol1 statement and all its options, include

```
symbol1;
```

before the new **symbol1** statement. To reset all the symbol definitions, include

```
goptions reset=symbol;
```

1.9.2 Overlaid Graphs

Overlaying two or more graphs is another technique that adds to the range of graphs that can be produced. The statement

```
plot y*x z*x / overlay ;
```

will produce a graph where **y** and **z** are both plotted against **x** on the same graph. Without the **overlay** option, two separate graphs would be produced. Chapter 8 has examples. Note that it is not possible to overlay graphs of the form **y*x=z**.

1.9.3 Viewing and Printing Graphics

For any program that produces graphics, we recommend beginning the program with

```
goptions reset=all;
```

and then setting all the options required explicitly. Under Microsoft Windows, a suitable set of graphics options might be:

```
goptions device=win target=winprtm rotate=landscape
ftext=swiss;
```

The **device=win** option specifies that the graphics are to be previewed on the screen. The **target=winprtm** option specifies that the hardcopy is to be produced on a monochrome printer set up in Windows, which can be configured from the **File**, **Print Setup** menu in SAS. For greyscale or colour printers, use **target=winprtg** or **target=winprtc**, respectively*.

The **rotate** option determines the orientation of the graphs. The alternative is **rotate=portrait**. The **ftext=swiss** option specifies a sans-serif font for the text in the graphs.

When a **goptions** statement such as this is used, the graphs will be displayed one by one in the graph window and the program will pause

* Under X-windows, the equivalent settings are **device=xcolor** and **target=xprintm**, **xprintg**, or **xprintc**.

between them with the message "Press Forward to see next graph" in the status line. The Page Down and Page Up keys are used for Forward and Backward, respectively.

1.10 Some Tips for Preventing and Correcting Errors

When writing programs:

1. One statement per line, where possible.
2. End each step with a run statement.
3. Indent each statement within a step (i.e., each statement between the data or proc statement and the run statement) by a couple of spaces. This is automated in the enhanced editor.
4. Give the full path name for raw data files on the infile statement.
5. Begin any programs that produce graphics with goptions reset=all; and then set the required options.

Before submitting a program:

1. Check that each statement ends with a semicolon.
2. Check that all opening and closing quotes match.

Use the enhanced editor colour coding to double-check.

3. Check any statement that does not begin with a keyword (blue, or navy blue) or a variable name (black).
4. Large blocks of purple may indicate a missing quotation mark.
5. Large areas of green may indicate a missing */ from a comment.

"Collapse" the program to check its overall structure. Hold down the Ctrl and Alt keys and press the numeric keypad minus key. Only the data, proc statements, and global statements should be visible. To expand the program, press the numeric keypad plus key while holding down Ctrl and Alt.

After running a program:

1. Examine the SAS log for warning and error messages.
2. Check for the message: "SAS went to a new line when INPUT statement reached past the end of a line" when using list input.
3. Verify that the number of observations and variables read in is correct.

4. Print out small data sets to ensure that they have been read correctly.

If there is an error message for a statement that appears to be correct, check whether the semicolon was omitted from the previous statement.

The message that a variable is "uninitialized" or "not found" usually means it has been misspelled.

To correct a missing quote, submit: '; run; or "; run; and then correct the program and resubmit it.

Chapter 2

Data Description and Simple Inference: Mortality and Water Hardness in the U.K.

2.1 Description of Data

The data to be considered in this chapter were collected in an investigation of environmental causes of diseases, and involve the annual mortality rates per 100,000 for males, averaged over the years from 1958 to 1964, and the calcium concentration (in parts per million) in the drinking water supply for 61 large towns in England and Wales. (The higher the calcium concentration, the harder the water.) The data appear in Table 7 of *SDS* and have been rearranged for use here as shown in Display 2.1. (Towns at least as far north as Derby are identified in the table by an asterisk.)

The main questions of interest about these data are as follows:

- How are mortality and water hardness related?
- Is there a geographical factor in the relationship?

2.2 Methods of Analysis

Initial examination of the data involves graphical techniques such as *histograms* and *normal probability plots* to assess the distributional properties of the two variables, to make general patterns in the data more visible, and to detect possible outliers. *Scatterplots* are used to explore the relationship between mortality and calcium concentration.

Following this initial graphical exploration, some type of correlation coefficient can be computed for mortality and calcium concentration. *Pearson's correlation coefficient* is generally used but others, for example, *Spearman's rank correlation*, may be more appropriate if the data are not considered to have a *bivariate normal distribution*. The relationship between the two variables is examined separately for northern and southern towns.

Finally, it is of interest to compare the mean mortality and mean calcium concentration in the north and south of the country by using either a *t-test* or its nonparametric alternative, the *Wilcoxon rank-sum test*.

2.3 Analysis Using SAS

Assuming the data is stored in an ASCII file, **water.dat**, as listed in Display 2.1 (i.e., including the '*' to identify the location of the town and the name of the town), then they can be read in using the following instructions:

	Town	*Mortality*	*Hardness*
	Bath	1247	105
*	Birkenhead	1668	17
	Birmingham	1466	5
*	Blackburn	1800	14
*	Blackpool	1609	18
*	Bolton	1558	10
*	Bootle	1807	15
	Bournemouth	1299	78
*	Bradford	1637	10
	Brighton	1359	84
	Bristol	1392	73
*	Burnley	1755	12
	Cardiff	1519	21
	Coventry	1307	78
	Croydon	1254	96
*	Darlington	1491	20
*	Derby	1555	39
*	Doncaster	1428	39

	East Ham	1318	122
	Exeter	1260	21
*	Gateshead	1723	44
*	Grimsby	1379	94
*	Halifax	1742	8
*	Huddersfield	1574	9
*	Hull	1569	91
	Ipswich	1096	138
*	Leeds	1591	16
	Leicester	1402	37
*	Liverpool	1772	15
*	Manchester	1828	8
*	Middlesbrough	1704	26
*	Newcastle	1702	44
	Newport	1581	14
	Northampton	1309	59
	Norwich	1259	133
*	Nottingham	1427	27
*	Oldham	1724	6
	Oxford	1175	107
	Plymouth	1486	5
	Portsmouth	1456	90
*	Preston	1696	6
	Reading	1236	101
*	Rochdale	1711	13
*	Rotherham	1444	14
*	St Helens	1591	49
*	Salford	1987	8
*	Sheffield	1495	14
	Southampton	1369	68
	Southend	1257	50
*	Southport	1587	75
*	South Shields	1713	71
*	Stockport	1557	13
*	Stoke	1640	57
*	Sunderland	1709	71
	Swansea	1625	13
*	Wallasey	1625	20
	Walsall	1527	60
	West Bromwich	1627	53
	West Ham	1486	122
	Wolverhampton	1485	81
*	York	1378	71

Display 2.1

```
data water;
   infile 'water.dat';
   input flag $ 1 Town $ 2-18 Mortal 19-22 Hardness 25-27;
   if flag = '*' then location = 'north';
      else location = 'south';
run;
```

The input statement uses SAS's column input where the exact columns containing the data for each variable are specified. Column input is simpler than list input in this case for three reasons:

- There is no space between the asterisk and the town name.
- Some town names are longer than eight characters — the default length for character variables.
- Some town names contain spaces, which would make list input complicated.

The univariate procedure can be used to examine the distributions of numeric variables. The following simple instructions lead to the results shown in Displays 2.2 and 2.3:

```
proc univariate data=water normal;
   var mortal hardness;
   histogram mortal hardness /normal;
   probplot mortal hardness;
run;
```

The normal option on the proc statement results in a test for the normality of the variables (see below). The var statement specifies which variable(s) are to be included. If the var statement is omitted, the default is *all* the numeric variables in the data set. The histogram statement produces histograms for both variables and the /normal option requests a normal distribution curve. Curves for various other distributions, including *non-parametric kernel density estimates* (see Silverman [1986]) can be produced by varying this option. Probability plots are requested with the probplot statement. *Normal probability plots* are the default. The resulting histograms and plots are shown in Displays 2.4 to 2.7.

Displays 2.2 and 2.3 provide significant information about the distributions of the two variables, mortality and hardness. Much of this is self-explanatory, for example, *Mean, Std Deviation, Variance*, and *N*. The meaning of some of the other statistics printed in these displays are as follows:

Uncorrected SS: Uncorrected sum of squares; simply the sum of squares of the observations

Corrected SS: Corrected sum of squares; simply the sum of squares of deviations of the observations from the sample mean

Coeff Variation: Coefficient of variation; the standard deviation divided by the mean and multiplied by 100

Std Error Mean: Standard deviation divided by the square root of the number of observations

Range: Difference between largest and smallest observation in the sample

Interquartile Range: Difference between the 25% and 75% quantiles (see values of quantiles given later in display to confirm)

Student's t: Student's *t*-test value for testing that the population mean is zero

Pr>|t|: Probability of a greater absolute value for the t-statistic

Sign Test: Nonparametric test statistic for testing whether the population median is zero

Pr>|M|: Approximation to the probability of a greater absolute value for the Sign test under the hypothesis that the population median is zero

Signed Rank: Nonparametric test statistic for testing whether the population mean is zero

Pr>=|S|: Approximation to the probability of a greater absolute value for the Sign Rank statistic under the hypothesis that the population mean is zero

Shapiro-Wilk W: Shapiro-Wilk statistic for assessing the normality of the data and the corresponding P-value (Shapiro and Wilk [1965])

Kolmogorov-Smirnov D: Kolmogorov-Smirnov statistic for assessing the normality of the data and the corresponding P-value (Fisher and Van Belle [1993])

Cramer-von Mises W-sq: Cramer-von Mises statistic for assessing the normality of the data and the associated P-value (Everitt [1998])

Anderson-Darling A-sq: Anderson-Darling statistic for assessing the normality of the data and the associated P-value (Everitt [1998])

The UNIVARIATE Procedure
Variable: Mortal

Moments

N	61	Sum Weights	61
Mean	1524.14754	Sum Observations	92973
Std Deviation	187.668754	Variance	35219.5612
Skewness	-0.0844436	Kurtosis	-0.4879484
Uncorrected SS	143817743	Corrected SS	2113173.67
Coeff Variation	12.3130307	Std Error Mean	24.0285217

Basic Statistical Measures

Location		Variability	
Mean	1524.148	Std Deviation	187.66875
Median	1555.000	Variance	35220
Mode	1486.000	Range	891.00000
		Interquartile Range	289.00000

NOTE: The mode displayed is the smallest of 3 modes with a count of 2.

Tests for Location: Mu0=0

Test	-Statistic-		-----P-value------	
Student's t	t	63.43077	Pr > |t|	<.0001
Sign	M	30.5	Pr >= |M|	<.0001
Signed Rank	S	945.5	Pr >= |S|	<.0001

Tests for Normality

Test	--Statistic---		-----P-value------	
Shapiro-Wilk	W	0.985543	Pr < W	0.6884
Kolmogorov-Smirnov	D	0.073488	Pr > D	>0.1500
Cramer-von Mises	W-Sq	0.048688	Pr > W-Sq	>0.2500
Anderson-Darling	A-Sq	0.337398	Pr > A-Sq	>0.2500

Quantiles (Definition 5)

Quantile	Estimate
100% Max	1987
99%	1987
95%	1800
90%	1742
75% Q3	1668
50% Median	1555
25% Q1	1379
10%	1259
5%	1247
1%	1096
0% Min	1096

Extreme Observations

----Lowest----		----Highest---	
Value	Obs	Value	Obs
1096	26	1772	29
1175	38	1800	4
1236	42	1807	7
1247	1	1828	30
1254	15	1987	46

Fitted Distribution for Mortal

Parameters for Normal Distribution

Parameter	Symbol	Estimate
Mean	Mu	1524.148
Std Dev	Sigma	187.6688

Goodness-of-Fit Tests for Normal Distribution

Test	---Statistic----		-----P-value-----	
Kolmogorov-Smirnov	D	0.07348799	Pr > D	>0.150
Cramer-von Mises	W-Sq	0.04868837	Pr > W-Sq	>0.250
Anderson-Darling	A-Sq	0.33739780	Pr > A-Sq	>0.250

Quantiles for Normal Distribution

	------Quantile------	
Percent	Observed	Estimated
1.0	1096.00	1087.56
5.0	1247.00	1215.46
10.0	1259.00	1283.64
25.0	1379.00	1397.57
50.0	1555.00	1524.15
75.0	1668.00	1650.73
90.0	1742.00	1764.65
95.0	1800.00	1832.84
99.0	1987.00	1960.73

Display 2.2

The UNIVARIATE Procedure

Variable: Hardness

Moments

N	61	Sum Weights	61
Mean	47.1803279	Sum Observations	2878
Std Deviation	38.0939664	Variance	1451.15027
Skewness	0.69223461	Kurtosis	-0.6657553
Uncorrected SS	222854	Corrected SS	87069.0164
Coeff Variation	80.7412074	Std Error Mean	4.8774326

Basic Statistical Measures

Location		Variability	
Mean	47.18033	Std Deviation	38.09397
Median	39.00000	Variance	1451
Mode	14.00000	Range	133.00000
		Interquartile Range	61.00000

Tests for Location: Mu0=0

Test	-Statistic-		-----P-value------	
Student's t	t	9.673189	Pr > |t|	<.0001
Sign	M	30.5	Pr >= |M|	<.0001
Signed Rank	S	945.5	Pr >= |S|	<.0001

Tests for Normality

Test	--Statistic---		-----P-value------	
Shapiro-Wilk	W	0.887867	Pr < W	<0.0001
Kolmogorov-Smirnov	D	0.196662	Pr > D	<0.0100
Cramer-von Mises	W-Sq	0.394005	Pr > W-Sq	<0.0050
Anderson-Darling	A-Sq	2.399601	Pr > A-Sq	<0.0050

Quantiles (Definition 5)

Quantile	Estimate
100% Max	138
99%	138
95%	122
90%	101
75% Q3	75
50% Median	39
25% Q1	14
10%	8
5%	6
1%	5
0% Min	5

Extreme Observations

----Lowest----		----Highest---	
Value	Obs	Value	Obs
5	39	107	38
5	3	122	19
6	41	122	59
6	37	133	35
8	46	138	26

Fitted Distribution for Hardness

Parameters for Normal Distribution

Parameter	Symbol	Estimate
Mean	Mu	47.18033
Std Dev	Sigma	38.09397

```
          Goodness-of-Fit Tests for Normal Distribution

Test                  ---Statistic----          -----P-value-----

Kolmogorov-Smirnov    D       0.19666241    Pr > D         <0.010
Cramer-von Mises      W-Sq    0.39400529    Pr > W-Sq      <0.005
Anderson-Darling      A-Sq    2.39960138    Pr > A-Sq      <0.005

                 Quantiles for Normal Distribution


                         --------Quantile-------
              Percent     Observed     Estimated

                  1.0      5.00000     -41.43949
                  5.0      6.00000     -15.47867
                 10.0      8.00000      -1.63905
                 25.0     14.00000      21.48634
                 50.0     39.00000      47.18033
                 75.0     75.00000      72.87432
                 90.0    101.00000      95.99971
                 95.0    122.00000     109.83933
                 99.0    138.00000     135.80015
```

Display 2.3

The quantiles provide information about the tails of the distribution as well as including the *five number summaries* for each variable. These consist of the minimum, lower quartile, median, upper quartile, and maximum values of the variables. The box plots that can be constructed from these summaries are often very useful in comparing distributions and identifying outliers. Examples are given in subsequent chapters.

The listing of extreme values can be useful for identifying outliers, especially when used with an id statement. The following section, entitled "Fitted Distribution for Hardness," gives details of the distribution fitted to the histogram. Because a normal distribution is fitted in this instance, it largely duplicates the output generated by the normal option on the proc statement.

The numerical information in Display 2.2 and the plots in Displays 2.4 and 2.5 all indicate that mortality is symmetrically, approximately normally, distributed. The formal tests of normality all result in non-significant values of the test statistic. The results in Display 2.3 and the plots in Displays 2.6 and 2.7, however, strongly suggest that calcium concentration (hardness) has a skew distribution with each of the tests for normality having associated P-values that are very small.

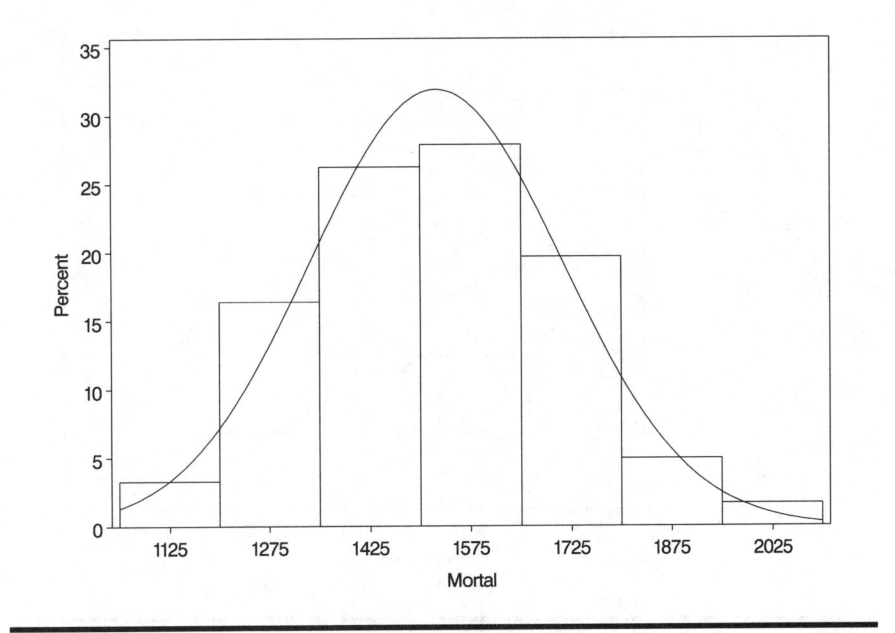

Display 2.4

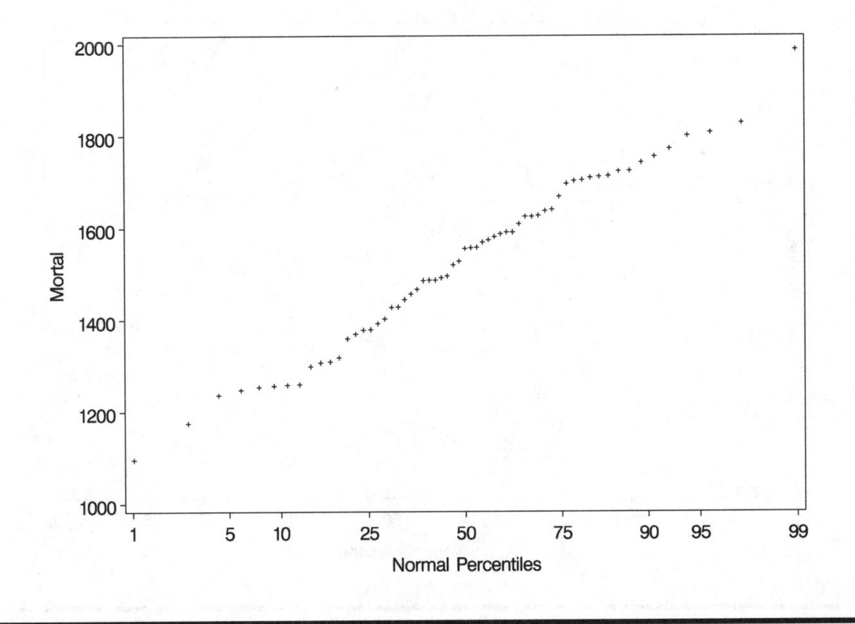

Display 2.5

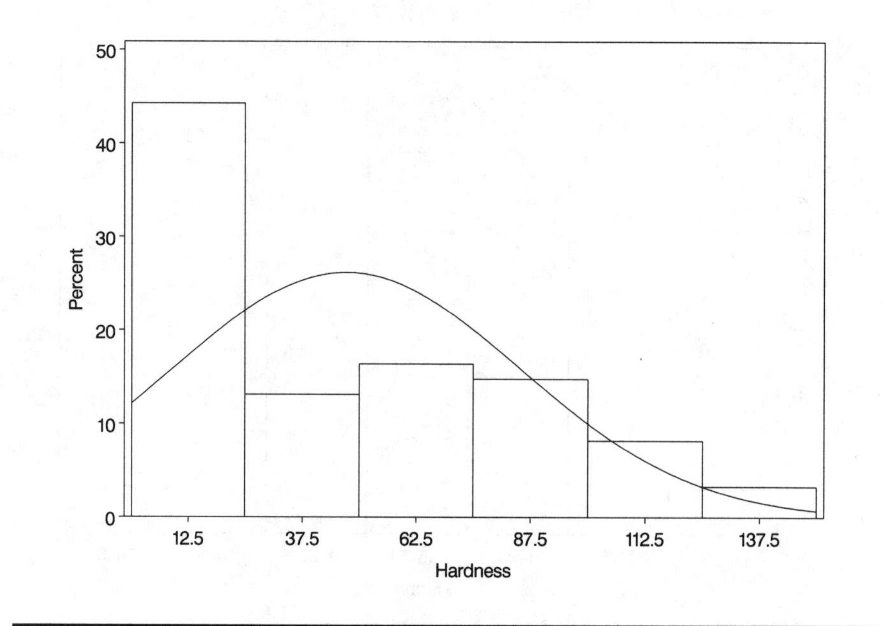

Display 2.6

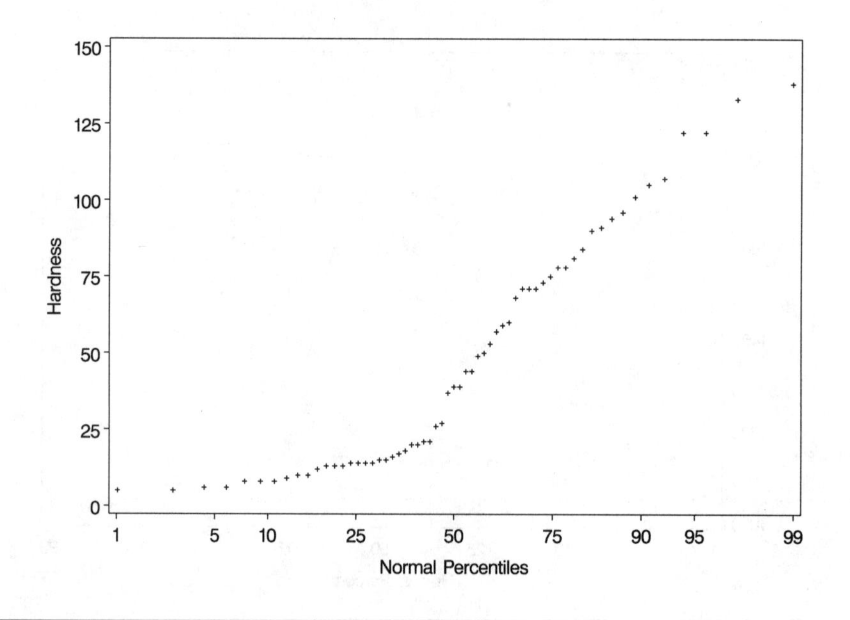

Display 2.7

The first step in examining the relationship between mortality and water hardness is to look at the scatterplot of the two variables. This can be found using **proc gplot** with the following instructions:

```
proc gplot;
   plot mortal*hardness;
run;
```

The resulting graph is shown in Display 2.8. The plot shows a clear negative association between the two variables, with high levels of calcium concentration tending to occur with low mortality values and vice versa. The correlation between the two variables is easily found using **proc corr**, with the following instructions:

```
proc corr data=water pearson spearman;
   var mortal hardness;
run;
```

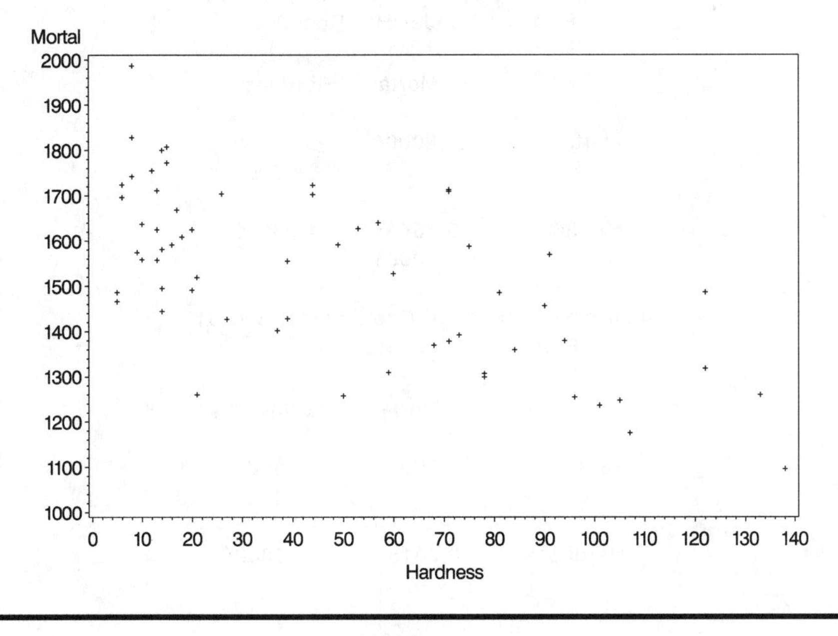

Display 2.8

The **pearson** and **spearman** options in the **proc corr** statement request that both types of correlation coefficient be calculated. The default, if neither option is used, is the Pearson coefficient.

The results from these instructions are shown in Display 2.9. The correlation is estimated to be –0.655 using the Pearson coefficient and –0.632 using Spearman's coefficient. In both cases, the test that the population correlation is zero has an associated P-value of 0.0001. There is clearly strong evidence for a non-zero correlation between the two variables.

```
The CORR Procedure

                    2 Variables:   Mortal   Hardness

                           Simple Statistics

Variable        N        Mean      Std Dev      Median    Minimum      Maximum

Mortal         61        1524    187.66875        1555       1096         1987
Hardness       61    47.18033     38.09397    39.00000    5.00000    138.00000

                Pearson Correlation Coefficients, N = 61
                      Prob > |r| under H0: Rho=0

                                Mortal    Hardness

                Mortal         1.00000    -0.65485
                                            <.0001

                Hardness      -0.65485     1.00000
                                <.0001

               Spearman Correlation Coefficients, N = 61
                      Prob > |r| under H0: Rho=0

                                Mortal    Hardness

                Mortal         1.00000    -0.63166
                                            <.0001

                Hardness      -0.63166     1.00000
                                <.0001
```

Display 2.9

One of the questions of interest about these data is whether or not there is a geographical factor in the relationship between mortality and water hardness, in particular whether this relationship differs between the towns in the North and those in the South. To examine this question, a useful first step is to replot the scatter diagram in Display 2.8 with northern and southern towns identified with different symbols. The necessary instructions are

```
symbol1 value=dot;
symbol2 value=circle;
proc gplot;
   plot mortal*hardness = location;
run;
```

The plot statement of the general form plot y * x = z will result in a scatter plot of y by x with a different symbol for each value of z. In this case, location has only two values and the first two plotting symbols used by SAS are 'x' and '+'. The symbol statements change the plotting symbols to give more impact to the scattergram.

The resulting plot is shown in Display 2.10. There appears to be no obvious difference in the form of the relationship between mortality and hardness for the two groups of towns.

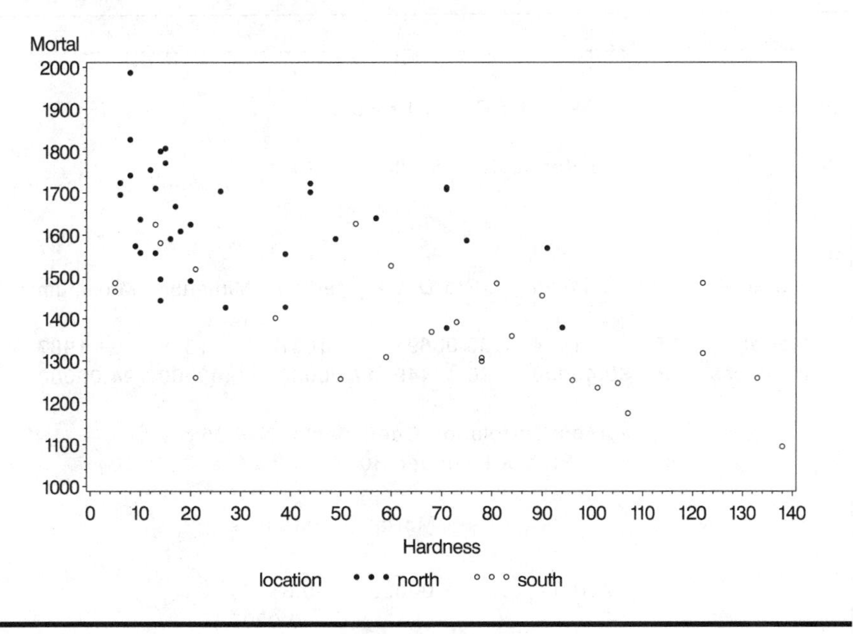

Display 2.10

Separate correlations for northern and southern towns can be produced using proc corr with a by statement as follows:

```
proc sort;
   by location;
proc corr data=water pearson spearman;
   var mortal hardness;
   by location;
run;
```

The by statement has the effect of producing separate analyses for each subgroup of the data defined by the specified variable, location in this case. However, the data set must first be sorted by that variable.

The results from this series of instructions are shown in Display 2.11. The main items of interest in this display are the correlation coefficients and the results of the tests that the population correlations are zero. The Pearson correlation for towns in the North is –0.369, and for those in the South it is –0.602. Both values are significant beyond the 5% level. The Pearson and Spearman coefficients take very similar values for this example.

```
------------------------------------ location=north -----------------------------------

                              The CORR Procedure

                       2 Variables:    Mortal    Hardness

                              Simple Statistics

Variable     N       Mean     Std Dev     Median   Minimum   Maximum

Mortal      35       1634   136.93691       1637      1378      1987
Hardness    35   30.40000    26.13449   17.00000   6.00000  94.00000

                  Pearson Correlation Coefficients, N = 35
                        Prob > |r| under H0: Rho=0

                                   Mortal    Hardness

                  Mortal          1.00000    -0.36860
                                               0.0293

                  Hardness       -0.36860     1.00000
                                   0.0293
```

```
            Spearman Correlation Coefficients, N = 35
                   Prob > |r| under H0: Rho=0

                              Mortal    Hardness

               Mortal        1.00000    -0.40421
                                          0.0160

               Hardness     -0.40421     1.00000
                              0.0160

------------------------------ location=south ------------------------------

                       The CORR Procedure

                 2 Variables:   Mortal   Hardness

                        Simple Statistics

Variable     N       Mean     Std Dev     Median     Minimum      Maximum

Mortal      26       1377   140.26918       1364        1096         1627
Hardness    26   69.76923    40.36068   75.50000     5.00000    138.00000

             Pearson Correlation Coefficients, N = 26
                   Prob > |r| under H0: Rho=0

                              Mortal    Hardness

               Mortal        1.00000    -0.60215
                                          0.0011

               Hardness     -0.60215     1.00000
                              0.0011

            Spearman Correlation Coefficients, N = 26
                   Prob > |r| under H0: Rho=0

                              Mortal    Hardness

               Mortal        1.00000    -0.59572
                                          0.0013

               Hardness     -0.59572     1.00000
                              0.0013
```

Display 2.11

Examination of scatterplots often centres on assessing density patterns such as clusters, gaps, or outliers. However, humans are not particularly good at visually examining point density and some type of density estimate added to the scatterplot is frequently very helpful. Here, plotting a *bivariate density estimate* for mortality and hardness is useful for gaining more insight into the structure of the data. (Details on how to calculate bivariate densities are given in Silverman [1986].) The following code produces and plots the bivariate density estimate of the two variables:

```
proc kde data=water out=bivest;
   var mortal hardness;
proc g3d data=bivest;
   plot hardness*mortal=density;
run;
```

The KDE procedure (proc kde) produces estimates of a univariate or bivariate probability density function using kernel density estimation (see Silverman [1986]). If a single variable is specified in the var statement, a univariate density is estimated and a bivariate density if two are specified. The out=bivest option directs the density estimates to a SAS data set. These can then be plotted with the three-dimensional plotting procedure proc g3d. The resulting plot is shown in Display 2.12. The two clear modes in the diagram correspond, at least approximately, to northern and southern towns.

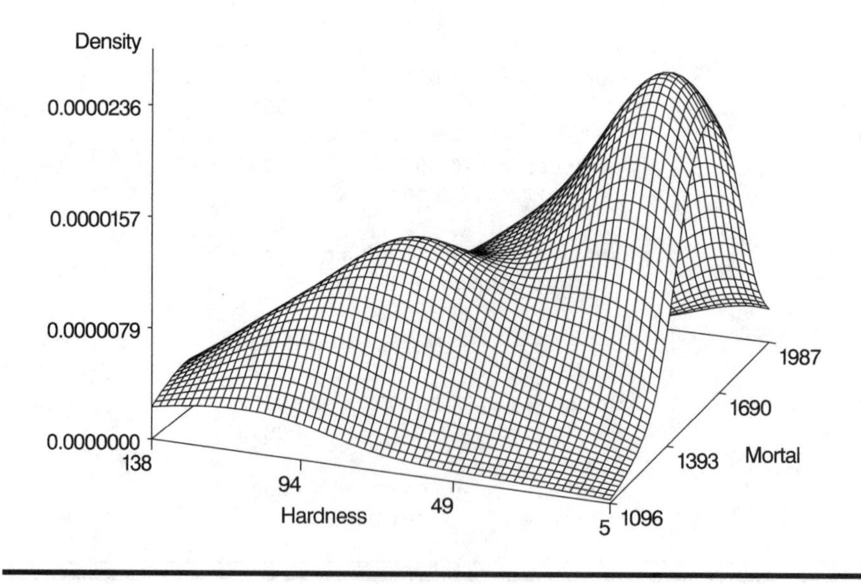

Display 2.12

The final question to address is whether or not mortality and calcium concentration differ in northern and southern towns. Because the distribution of mortality appears to be approximately normal, a *t*-test can be applied. Calcium concentration has a relatively high degree of skewness; thus, applying a Wilcoxon test or a *t*-test after a log transformation may be more sensible. The relevant SAS instructions are as follows.

```
data water;
   set water;
   lhardnes=log(hardness);
proc ttest;
   class location;
   var mortal hardness lhardnes;
proc npar1way wilcoxon;
   class location;
   var hardness;
run;
```

The short data step computes the (natural) log of **hardness** and stores it in the data set as the variable **lhardnes.** To use **proc ttest**, the variable that divides the data into two groups is specified in the **class** statement and the variable (or variables) whose means are to be compared are specified in the **var** statement. For a Wilcoxon test, the **npar1way** procedure is used with the **wilcoxon** option.

The results of the *t*-tests are shown in Display 2.13; those for the Wilcoxon tests in Display 2.14. The *t*-test for mortality gives very strong evidence for a difference in mortality in the two regions, with that in the North being considerably larger (the 95% confidence interval for the difference is 185.11, 328.47). Using a test that assumes equal variances in the two populations or one that does not make this assumption (Satterthwaite [1946]) makes little difference in this case. The *t*-test on the untransformed hardness variable also indicates a difference, with the mean hardness in the North being far less than amongst towns in the South. Notice here that the test for the equality of population variances (one of the assumptions of the *t*-test) suggests that the variances differ. In examining the results for the log-transformed variable, it is seen that the *t*-test still indicates a highly significant difference, but in this case the test for homogeneity is nonsignificant.

The result from the nonparametric Wilcoxon test (Display 2.14) once again indicates that the mean water hardness of towns in the North differs from that of towns in the South.

The TTEST Procedure

Statistics

Variable	Class	N	Lower CL Mean	Mean	Upper CL Mean	Lower CL Std Dev	Std Dev	Upper CL Std Dev	Std Err
Mortal	north	35	1586.6	1633.6	1680.6	110.76	136.94	179.42	23.147
Mortal	south	26	1320.2	1376.8	1433.5	110.01	140.27	193.63	27.509
Mortal	Diff (1-2)		185.11	256.79	328.47	117.28	138.36	168.75	35.822
Hardness	north	35	21.422	30.4	39.378	21.139	26.134	34.241	4.4175
Hardness	south	26	53.467	69.769	86.071	31.653	40.361	55.714	7.9154
Hardness	Diff (1-2)		-56.43	-39.37	-22.31	27.906	32.922	40.154	8.5237
Lhardnes	north	35	2.7887	3.0744	3.3601	0.6727	0.8316	1.0896	0.1406
Lhardnes	south	26	3.5629	3.9484	4.3339	0.7485	0.9544	1.3175	0.1872
Lhardnes	Diff (1-2)		-1.333	-0.874	-0.415	0.7508	0.8857	1.0803	0.2293

T-Tests

Variable	Method	Variances	DF	t Value	Pr > |t|
Mortal	Pooled	Equal	59	7.17	<.0001
Mortal	Satterthwaite	Unequal	53.3	7.14	<.0001
Hardness	Pooled	Equal	59	-4.62	<.0001
Hardness	Satterthwaite	Unequal	40.1	-4.34	<.0001
lhardnes	Pooled	Equal	59	-3.81	0.0003
lhardnes	Satterthwaite	Unequal	49.6	-3.73	0.0005

Equality of Variances

Variable	Method	Num DF	Den DF	F Value	Pr > F
Mortal	Folded F	25	34	1.05	0.8830
Hardness	Folded F	25	34	2.39	0.0189
lhardnes	Folded F	25	34	1.32	0.4496

Display 2.13

The NPAR1WAY Procedure

Wilcoxon Scores (Rank Sums) for Variable Hardness
Classified by Variable location

location	N	Sum of Scores	Expected Under H0	Std Dev Under H0	Mean Score
north	35	832.50	1085.0	68.539686	23.785714
south	26	1058.50	806.0	68.539686	40.711538

Average scores were used for ties.

```
                 Wilcoxon Two-Sample Test

         Statistic                   1058.5000

         Normal Approximation
         Z                              3.6767
         One-Sided Pr > Z               0.0001
         Two-Sided Pr > |Z|             0.0002

         t Approximation
         One-Sided Pr > Z               0.0003
         Two-Sided Pr > |Z|             0.0005

       Z includes a continuity correction of 0.5.

                  Kruskal-Wallis Test

           Chi-Square               13.5718
           DF                             1
           Pr > Chi-Square           0.0002
```

Display 2.14

Exercises

2.1 Rerun **proc univariate** with the **plot** option for line printer plots.

2.2 Generate box plots of mortality and water hardness by location (use **proc boxplot**).

2.3 Use **proc univariate** to compare the distribution of water hardness to the log normal and exponential distributions.

2.4 Produce histograms of both mortality and water hardness with, in each case, a kernel density estimate of the variable's distribution superimposed.

2.5 Produce separate perspective plots of the estimated bivariate densities of northern and southern towns.

2.6 Reproduce the scatterplot in Display 2.10 with added linear regression fits of mortality and hardness for both northern and southern towns. Use different line types for the two regions.

Chapter 3

Simple Inference for Categorical Data: From Sandflies to Organic Particulates in the Air

3.1 Description of Data

This chapter considers the analysis of categorical data. It begins by looking at tabulating raw data into cross-classifications (i.e. *contingency tables*) using the mortality and water hardness data from Chapter 2. It then examines six data sets in which the data are already tabulated; a description of each of these data sets is given below. The primary question of interest in each case involves assessing the relationship between pairs of categorical variables using the chi-square test or some suitable alternative.

The six cross-classified data sets to be examined in this chapter are as follows:

1. **Sandflies (Table 128 in *SDS*)**. These data are given in Display 3.1 and show the number of male and female sandflies caught in light traps set 3 ft and 35 ft above the ground at a site in eastern Panama. The question of interest is: does the proportion of males and females caught at a particular height differ?

	Sandflies	
	3 ft	*35 ft*
Males	173	125
Females	150	73
Total	323	198

Display 3.1

2. **Acacia ants (Table 27 in *SDS*)**. These data, given in Display 3.2, record the results of an experiment with acacia ants. All but 28 trees of two species of acacia were cleared from an area in Central America, and the 28 trees were cleared of ants using insecticide. Sixteen colonies of a particular species of ant were obtained from other trees of species A. The colonies were placed roughly equidistant from the 28 trees and allowed to invade them. The question of interest is whether the invasion rate differs for the two species of acacia tree.

Acacia Species	*Not Invaded*	*Invaded*	*Total*
A	2	13	15
B	10	3	13
Total	12	16	28

Display 3.2

3. **Piston ring failures (Table 15 in *SDS*)**. These data are reproduced in Display 3.3 and show the number of failures of piston rings in each of three legs in each of four steam-driven compressors located in the same building. The compressors have identical design and are orientated in the same way. The question of interest is whether the pattern of the location of failures is different for different compressors.
4. **Oral contraceptives**. These data appear in Display 3.4 and arise from a study reported by Sartwell et al. (1969). The study was

conducted in a number of hospitals in several large American cities. In those hospitals, all those married women identified as suffering from idiopathic thromboembolism (blood clots) over a 3-year period were individually matched with a suitable control, those being female patients discharged alive from the same hospital in the same 6-month time interval as the case. In addition, they were individually matched to cases on age, marital status, race, etc. Patients and controls were then asked about their use of oral contraceptives.

Compressor No.	*North*	*Centre*	*South*	*Total*
1	17	17	12	46
2	11	9	13	33
3	11	8	19	38
4	14	7	28	49
Total	53	41	72	166

Display 3.3

	Controls	
Oral Contraceptive Use	*Used*	*Not Used*
Cases used	10	57
Cases not used	13	95

Display 3.4

5. **Oral lesions**. These data appear in Display 3.5; they give the location of oral lesions obtained in house-to-house surveys in three geographic regions of rural India.
6. **Particulates in the air**. These data are given in Display 3.6; they arise from a study involving cases of bronchitis by level of organic particulates in the air and by age (Somes and O'Brien [1985]).

	Region		
Site of Lesion	*Keral*	*Gujarat*	*Andhra*
Buccal mucosa	8	1	8
Labial mucosa	0	1	0
Commissure	0	1	0
Gingiva	0	1	0
Hard palate	0	1	0
Soft palate	0	1	0
Tongue	0	1	0
Floor of mouth	1	0	1
Alveolar ridge	1	0	1

Display 3.5 (Taken from Table 18.1 of the *User Manual for Proc.-Stat Xact-4 for SAS Users,* Cytel Software Corporation, Cambridge, MA, 2000.)

		Bronchitis		
Age (Years)	*Particulates Level*	*Yes*	*No*	*Total*
15–24	High	20	382	402
	Low	9	214	223
23–30	High	10	172	182
	Low	7	120	127
40+	High	12	327	339
	Low	6	183	189

Display 3.6 (Reprinted by permission from the *Encyclopedia of Statistical Sciences*, John Wiley & Sons, Inc., New York, Copyright © 1982.)

3.2 Methods of Analysis

Contingency tables are one of the most common ways to summarize categorical data. Displays 3.1, 3.2, and 3.4 are examples of 2 × 2 contingency tables (although Display 3.4 has a quite different structure from Displays

3.1 and 3.2 as explained later). Display 3.3 is an example of a 3 × 4 table and Display 3.5 an example of a 9 × 3 table with very sparse data. Display 3.6 is an example of a series of 2 × 2 tables involving the same two variables. For all such tables, interest generally lies in assessing whether or not there is an association between the row variable and the column variable that form the table. Most commonly, a *chi-square test of independence* is used to answer this question, although alternatives such as *Fisher's exact test* or *McNemar's test* may be needed when the sample size is small (Fisher's test) or the data consists of matched samples (McNemar's test). In addition, in 2 × 2 tables, it may be required to calculate a confidence interval for the difference in two population proportions. For a series of 2 × 2 tables, the *Mantel-Haenszel test* may be appropriate (see later). (Details of all the tests mentioned are given in Everitt [1992].)

3.3 Analysis Using SAS

3.3.1 Cross-Classifying Raw Data

We first demonstrate how raw data can be put into the form of a cross-classification using the data on mortality and water hardness from Chapter 2.

```
data water;
   infile 'n:\handbook\datasets\water.dat';
   input flag $ 1 Town $ 2-18 Mortal 19-22 Hardness 25-27;
   if flag='*' then location='north';
      else location='south';
   mortgrp=mortal > 1555;
   hardgrp=hardness > 39;
run;
proc freq data=water;
   tables mortgrp*hardgrp / chisq;
run;
```

The raw data are read into a SAS data set water, as described in Chapter 2. In this instance, two new variables are computed — mortgrp and hardgrp — that dichotomise mortality and water hardness at their medians. Strictly speaking, the expression mortal > 1555 is a logical expression yielding the result "true" or "false," but in SAS these are represented by the values 1 and 0, respectively.

Proc freq is used both to produce contingency tables and to analyse them. The tables statement defines the table to be produced and specifies the analysis of it. The variables that form the rows and columns are joined with an asterisk (*); these may be numeric or character variables. One-way frequency distributions are produced where variables are not joined by asterisks. Several tables can be specified in a single **tables** statement.

The options after the "/" specify the type of analysis. The **chisq** option requests chi-square tests of independence and measures of association based on chi-square. The output is shown in Display 3.7. We leave commenting on the contents of this type of output until later.

The FREQ Procedure

Table of mortgrp by hardgrp

mortgrp	hardgrp		
Frequency Percent Row Pct Col Pct	0	1	Total
0	11 8.03 35.48 34.38	20 32.79 64.52 68.97	31 50.82
1	21 34.43 70.00 65.63	9 14.75 30.00 31.03	30 49.18
Total	32 52.46	29 47.54	61 100.00

Statistics for Table of mortgrp by hardgrp

Statistic	DF	Value	Prob
Chi-Square	1	7.2830	0.0070
Likelihood Ratio Chi-Square	1	7.4403	0.0064
Continuity Adj. Chi-Square	1	5.9647	0.0146
Mantel-Haenszel Chi-Square	1	7.1636	0.0074
Phi Coefficient		-0.3455	
Contingency Coefficient		0.3266	
Cramer's V		-0.3455	

```
                Fisher's Exact Test
        ------------------------------------------
        Cell (1,1) Frequency (F)          11
        Left-sided Pr <= F            0.0070
        Right-sided Pr >= F           0.9986
        Table Probability (P)         0.0056
        Two-sided Pr <= P             0.0103

                 Sample Size = 61
```

Display 3.7

Now we move on to consider the six data sets that actually arise in the form of contingency tables. The freq procedure is again used to analyse such tables and compute tests and measures of association.

3.3.2 Sandflies

The data on sandflies in Display 3.1. can be read into a SAS data set with each cell as a separate observation and the rows and columns identified as follows:

```
data sandflies;
   input sex $ height n;
cards;
m 3 173
m 35 125
f 3 150
f 35 73
;
```

The rows are identified by a character variable sex with values m and f. The columns are identifed by the variable height with values 3 and 35. The variable n contains the cell count. proc freq can then be used to analyse the table.

```
proc freq data=sandflies;
   tables sex*height /chisq riskdiff;
   weight n;
run;
```

The **riskdiff** option requests differences in risks (or binomial proportions) and their confidence limits.

The **weight** statement specifies a variable that contains weights for each observation. It is most commonly used to specify cell counts, as in this example. The default weight is 1, so the **weight** statement is not required when the data set consists of observations on individuals.

The results are shown in Display 3.8. First, the 2 × 2 table of data is printed, augmented with total frequency, row, and column percentages. A number of statistics calculated from the frequencies in the table are then printed, beginning with the well-known chi-square statistic used to test for the independence of the two variables forming the table. Here, the P-value associated with the chi-square statistic suggests that sex and height are not independent. The *likelihood ratio chi-square* is an alternative statistic for testing independence (as described in Everitt [1992]). Here, the usual chi-square statistic and the likelihood ratio statistic take very similar values. Next the continuity adjusted chi-square statistic is printed. This involves what is usually known as *Yates's correction*, again described in Everitt (1992). The correction is usually suggested as a simple way of dealing with what was once perceived as the problem of unacceptably small frequencies in contingency tables. Currently, as seen later, there are much better ways of dealing with the problem and really no need to ever use Yates's correction. The Mantel-Haenszel statistic tests the hypothesis of a linear association between the row and column variables. Both should be ordinal numbers. The next three statistics printed in Display 3.8 — namely, the *Phi coefficient*, *Contingency coefficient*, and *Cramer's V* — are all essentially attempts to quantify the degree of the relationship between the two variables forming the contingency table. They are all described in detail in Everitt (1992).

Following these statistics in Display 3.8 is information on Fisher's exact test. This is more relevant to the data in Display 3.2 and thus its discussion comes later. Next come the results of estimating a confidence interval for the difference in proportions in the contingency table. Thus, for example, the estimated difference in the proportion of female and male sandflies caught in the 3-ft light traps is 0.0921 (0.6726–0.5805). The standard error of this difference is calculated as:

$$\sqrt{\frac{0.6726(1-0.6726)}{223}+\frac{0.5805(1-0.5805)}{298}}$$

that is, the value of 0.0425 given in Display 3.8. The confidence interval for the difference in proportions is therefore:

$$0.0921 \pm 1.96 \times 0.0425 = (0.0088,\ 0.1754)$$

as given in Display 3.8. The proportion of female sandflies caught in the 3-ft traps is larger than the corresponding proportion for males.

The FREQ Procedure

Table of sex by height

Sex / Frequency, Percent, Row Pct, Col Pct	height 3	35	Total
f	150 28.79 67.26 46.44	73 14.01 32.74 36.87	223 42.80
m	173 33.21 58.05 53.56	125 23.99 41.95 63.13	298 57.20
Total	323 62.00	198 38.00	521 100.00

Statistics for Table of sex by height

Statistic	DF	Value	Prob
Chi-Square	1	4.5930	0.0321
Likelihood Ratio Chi-Square	1	4.6231	0.0315
Continuity Adj. Chi-Square	1	4.2104	0.0402
Mantel-Haenszel Chi-Square	1	4.5842	0.0323
Phi Coefficient		0.0939	
Contingency Coefficient		0.0935	
Cramer's V		0.0939	

Fisher's Exact Test

Cell (1,1) Frequency (F)	150
Left-sided Pr <= F	0.9875
Right-sided Pr >= F	0.0199
Table Probability (P)	0.0073
Two-sided Pr <= P	0.0360

Column 1 Risk Estimates

	Risk	ASE	(Asymptotic) 95% Confidence Limits		(Exact) 95% Confidence Limits	
Row 1	0.6726	0.0314	0.6111	0.7342	0.6068	0.7338
Row 2	0.5805	0.0286	0.5245	0.6366	0.5223	0.6372
Total	0.6200	0.0213	0.5783	0.6616	0.5767	0.6618
Difference	0.0921	0.0425	0.0088	0.1754		

Difference is (Row 1 – Row 2)

Column 2 Risk Estimates

	Risk	ASE	(Asymptotic) 95% Confidence Limits		(Exact) 95% Confidence Limits	
Row 1	0.3274	0.0314	0.2658	0.3889	0.2662	0.3932
Row 2	0.4195	0.0286	0.3634	0.4755	0.3628	0.4777
Total	0.3800	0.0213	0.3384	0.4217	0.3382	0.4233
Difference	-0.0921	0.0425	-0.1754	-0.0088		

Difference is (Row 1 - Row 2)

Sample Size = 521

Display 3.8

3.3.3 Acacia Ants

The acacia ants data are also in the form of a contingency table and are read in as four observations representing cell counts.

```
data ants;
   input species $ invaded $ n;
cards;
A no 2
A yes 13
```

```
B no 10
B no 3
;
proc freq data=ants;
   tables species*invaded / chisq expected;
   weight n;
run;
```

In this example, the **expected** option in the **tables** statement is used to print expected values under the independence hypothesis for each cell.

The results are shown in Display 3.9. Here, because of the small frequencies in the table, Fisher's exact test might be the preferred option, although all the tests of independence have very small associated P-values and thus, very clearly, species and invasion are not independent. A higher proportion of ants invaded species A than species B.

```
The FREQ Procedure

Table of species by invaded

species      invaded

Frequency|
Expected |
Percent  |
Row Pct  |
Col Pct  |no      |yes     |  Total
---------+--------+--------+
A        |      2 |     13 |     15
         | 8.0357 | 6.9643 |
         |   7.14 |  46.43 |  53.57
         |  13.33 |  86.67 |
         |  13.33 | 100.00 |
---------+--------+--------+
B        |     13 |      0 |     13
         | 6.9643 | 6.0357 |
         |  46.43 |   0.00 |  46.43
         | 100.00 |   0.00 |
         |  86.67 |   0.00 |
---------+--------+--------+
Total          15       13       28
            53.57    46.43   100.00
```

Statistics for Table of species by invaded

Statistic	DF	Value	Prob
Chi-Square	1	21.0311	<.0001
Likelihood Ratio Chi-Square	1	26.8930	<.0001
Continuity Adj. Chi-Square	1	17.6910	<.0001
Mantel-Haenszel Chi-Square	1	20.2800	<.0001
Phi Coefficient		-0.8667	
Contingency Coefficient		0.6549	
Cramer's V		-0.8667	

Fisher's Exact Test

Cell (1,1) Frequency (F)	2
Left-sided Pr <= F	2.804E-06
Right-sided Pr >= F	1.0000
Table Probability (P)	1.0000
Two-sided Pr <= P	2.831E-06

Sample Size = 28

Display 3.9

3.3.4 *Piston Rings*

Moving on to the piston rings data, they are read in and analysed as follows:

```
data pistons;
   input machine site $ n;
cards;
1 North 17
1 Centre 17
```

```
1 South 12
2 North 11
2 Centre 9
2 South 13
3 North 11
3 Centre 8
3 South 19
4 North 14
4 Centre 7
4 South 28
;

proc freq data=pistons order=data;
   tables machine*site / chisq deviation cellchi2 norow nocol
nopercent;
   weight n;
run;
```

The **order=data** option in the **proc** statement specifies that the rows and columns of the tables follow the order in which they occur in the data. The default is number order for numeric variables and alphabetical order for character variables.

The **deviation** option in the **tables** statement requests the printing of residuals in the cells, and the **cellchi2** option requests that each cell's contribution to the overall chi-square be printed. To make it easier to view the results the, row, column, and overall percentages are suppressed with the **norow**, **nocol**, and **nopercent** options, respectively.

Here, the chi-square test for independence given in Display 3.10 shows only relatively weak evidence of a departure from independence. (The relevant P-value is 0.069). However, the simple residuals (the differences between an observed frequency and that expected under independence) suggest that failures are fewer than might be expected in the South leg of Machine 1 and more than expected in the South leg of Machine 4. (Other types of residuals may be more useful — see Exercise 3.2.)

The FREQ Procedure

Table of machine by site

Machine / Frequency / Deviation / Cell Chi-Square	Site: North	Centre	South	Total
1	7 2.3133 0.3644	17 5.6386 2.7983	12 -7.952 3.1692	46
2	11 0.4639 0.0204	9 0.8494 0.0885	13 -1.313 0.1205	33
3	11 -1.133 0.1057	8 -1.386 0.2045	19 2.5181 0.3847	38
4	4 -1.645 0.1729	7 -5.102 2.1512	28 6.747 2.1419	49
Total	53	41	72	166

Statistics for Table of machine by site

Statistic	DF	Value	Prob
Chi-Square	6	11.7223	0.0685
Likelihood Ratio Chi-Square	6	12.0587	0.0607
Mantel-Haenszel Chi-Square	1	5.4757	0.0193
Phi Coefficient		0.2657	
Contingency Coefficient		0.2568	
Cramer's V		0.1879	

Sample Size = 166

Display 3.10

3.3.5 Oral Contraceptives

The oral contraceptives data involve *matched* observations. Consequently, they cannot be analysed with the usual chi-square statistic. Instead, they require application of McNemar's test, as described in Everitt (1992). The data can be read in and analysed with the following SAS commands:

```
data the_pill;
   input caseuse $ contruse $ n;
cards;
Y Y 10
Y N 57
N Y 13
N N 95
;
proc freq data=the_pill order=data;
   tables caseuse*contruse / agree;
   weight n;
run;
```

The **agree** option on the tables statement requests measures of agreement, including the one of most interest here, the McNemar test. The results appear in Display 3.11. The test of no association between using oral contraceptives and suffering from blood clots is rejected. The proportion of matched pairs in which the case has used oral contraceptives and the control has not is considerably higher than pairs where the reverse is the case.

The FREQ Procedure

Table of caseuse by contruse

caseuse Frequency Percent Row Pct Col Pct	contruse Y	N	Total
Y	10 5.71 14.93 43.48	57 32.57 85.07 37.50	67 38.29
N	13 7.43 12.04 56.52	95 54.29 87.96 62.50	108 61.71
Total	23 13.14	152 86.86	175 100.00

```
Statistics for Table of caseuse by contruse

              McNemar's Test
        ------------------------
        Statistic (S)    27.6571
        DF                     1
        Pr > S            <.0001

         Simple Kappa Coefficient
    --------------------------------
    Kappa                    0.0330
    ASE                      0.0612
    95% Lower Conf Limit    -0.0870
    95% Upper Conf Limit     0.1530

            Sample Size = 175
```

Display 3.11

3.3.6 Oral Cancers

The data on the regional distribution of oral cancers is read in using the following data step:

```
data lesions;
   length region $8.;
   input site $ 1-16 n1 n2 n3;
   region='Keral';
   n=n1;
   output;
   region='Gujarat';
   n=n2;
   output;
   region='Anhara';
   n=n3;
   output;
   drop n1-n3;
   cards;
   Buccal Mucosa  8   1   8
   Labial Mucosa  0   1   0
```

```
Commissure      0  1  0
Gingiva         0  1  0
Hard palate     0  1  0
Soft palate     0  1  0
Tongue          0  1  0
Floor of mouth  1  0  1
Alveolar ridge  1  0  1
;
```

This data step reads in the values for three cell counts from a single line of instream data and then creates three separate observations in the output data set. This is achieved using three **output** statements in the data step. The effect of each **output** statement is to create an observation in the data set with the data values that are current at that point. First, the **input** statement reads a line of data that contains the three cell counts. It uses column input to read the first 16 columns into the **site** variable, and then the list input to read the three cell counts into variables **n1** to **n3**. When the first **output** statement is executed, the **region** variable has been assigned the value **'Keral'** and the variable **n** has been set equal to the first of the three cell counts read in. At the second **output** statement, the value of **region** is **'Gujarat'**, and **n** equals **n2**, the second cell count, and so on for the third **output** statement. When restructuring data like this, it is wise to check the results, either by viewing the resultant data set interactively or using **proc print**. The SAS log also gives the number of variables and observations in any data set created and thus can be used to provide a check.

The **drop** statement excludes the variables mentioned from the lesions data set.

```
proc freq data=lesions order=data;
   tables site*region /exact;
   weight n;
run;
```

For 2 × 2 tables, Fisher's exact test is calculated and printed by default. For larger tables, exact tests must be explicitly requested with the **exact** option on the **tables** statement. Here, because of the very sparse nature of the data, it is likely that the exact approach will differ from the usual chi-square procedure. The results given in Display 3.12 confirm this. The chi-square test has an associated P-value of 0.14, indicating that the hypothesis of independence site and region is acceptable. The exact test has an associated P-value of 0.01, indicating that the site of lesion and region are associated. Here, the chi-square test is unsuitable because of the very sparse nature of the data.

```
The FREQ Procedure
                    Table of site by region

site                        region

Frequency
Percent
Row Pct
Col Pct        | Keral  | Gujarat | Anhara |  Total
---------------------------------------------------
Buccal Mucosa  |     8  |      1  |     8  |     17
               | 29.63  |   3.70  | 29.63  |  62.96
               | 47.06  |   5.88  | 47.06  |
               | 80.00  |  14.29  | 80.00  |
---------------------------------------------------
Labial Mucosa  |     0  |      1  |     0  |      1
               |  0.00  |   3.70  |  0.00  |   3.70
               |  0.00  | 100.00  |  0.00  |
               |  0.00  |  14.29  |  0.00  |
---------------------------------------------------
Commissure     |     0  |      1  |     0  |      1
               |  0.00  |   3.70  |  0.00  |   3.70
               |  0.00  | 100.00  |  0.00  |
               |  0.00  |  14.29  |  0.00  |
---------------------------------------------------
Gingiva        |     0  |      1  |     0  |      1
               |  0.00  |   3.70  |  0.00  |   3.70
               |  0.00  | 100.00  |  0.00  |
               |  0.00  |  14.29  |  0.00  |
---------------------------------------------------
Hard palate    |     0  |      1  |     0  |      1
               |  0.00  |   3.70  |  0.00  |   3.70
               |  0.00  | 100.00  |  0.00  |
               |  0.00  |  14.29  |  0.00  |
---------------------------------------------------
Soft palate    |     0  |      1  |     0  |      1
               |  0.00  |   3.70  |  0.00  |   3.70
               |  0.00  | 100.00  |  0.00  |
               |  0.00  |  14.29  |  0.00  |
---------------------------------------------------
Tongue         |     0  |      1  |     0  |      1
               |  0.00  |   3.70  |  0.00  |   3.70
               |  0.00  | 100.00  |  0.00  |
               |  0.00  |  14.29  |  0.00  |
---------------------------------------------------
```

Floor of mouth	1	0	1	2
	3.70	0.00	3.70	7.41
	50.00	0.00	50.00	
	10.00	0.00	10.00	
Alveolar ridge	1	0	1	2
	3.70	0.00	3.70	7.41
	50.00	0.00	50.00	
	10.00	0.00	10.00	
Total	10	7	10	27
	37.04	25.93	37.04	100.00

Statistics for Table of site by region

Statistic	DF	Value	Prob
Chi-Square	16	22.0992	0.1400
Likelihood Ratio Chi-Square	16	23.2967	0.1060
Mantel-Haenszel Chi-Square	1	0.0000	1.0000
Phi Coefficient		0.9047	
Contingency Coefficient		0.6709	
Cramer's V		0.6397	

WARNING: 93% of the cells have expected counts less than 5. Chi-Square may not be a valid test.

The FREQ Procedure

Statistics for Table of site by region

Fisher's Exact Test

Table Probability (P)	5.334E-06
Pr <=P	0.0101

Sample Size = 27

Display 3.12

3.3.7 Particulates and Bronchitis

The final data set to be analysed in this chapter, the bronchitis data in Display 3.6, involves 2 × 2 tables for bronchitis and level of organic particulates for three age groups. The data could be collapsed over age and the aggregate 2 × 2 table analysed as described previously. However, the potential dangers of this procedure are well-documented (see, for example, Everitt [1992]). In particular, such pooling of contingency tables can generate an association when in the separate tables there is none. A more appropriate test in this situation is the Mantel-Haenszel test. For a series of k 2 × 2 tables, the test statistic for testing the hypothesis of no association is:

$$X^2 = \frac{\left[\sum_{i=1}^{k} a_i - \sum_{i=1}^{k} \frac{(a_i + b_i)(a_i + c_i)}{n_i}\right]^2}{\sum_{i=1}^{k} \frac{(a_i + b_i)(c_i + d_i)(a_i + c_i)(b_i + d_i)}{n_i^2(n_i - 1)}} \tag{3.1}$$

where a_i, b_i, c_i, d_i represent the counts in the four cells of the ith table and n_i is the total number of observations in the ith table. Under the null hypothesis of independence in all tables, this statistic has a chi-squared distribution with a single degree of freedom.

The data can be read in and analysed using the following SAS code:

```
data bronchitis;
   input agegrp level $ bronch $ n;
cards;
1     H  Y  20
1     H  N  382
1     L  Y  9
1     L  N  214
2     H  Y  10
2     H  N  172
2     L  Y  7
2     L  N  120
3     H  Y  12
3     H  N  327
3     L  Y  6
3     L  N  183
;
proc freq data=bronchitis order=data;
   Tables agegrp*level*bronch / cmh noprint;
   weight n;
run;
```

The **tables** statement specifies a three-way tabulation with **agegrp** defining the strata. The **cmh** option requests the Cochran-Mantel-Haenszel statistics and the **noprint** option suppresses the tables. The results are shown in Display 3.13. There is no evidence of an association between level of organic particulates and suffering from bronchitis. The P-value associated with the test statistic is 0.64 and the assumed common odds ratio calculated as:

$$\hat{\psi}_{\text{pooled}} = \frac{\Sigma(a_i d_i / n_i)}{\Sigma(b_i c_i / n_i)} \tag{3.2}$$

takes the value 1.13 with a confidence interval of 0.67, 1.93. (Since the Mantel-Haenszel test will only give sensible results if the association between the two variables is both the same size and same direction in each 2 × 2 table, it is generally sensible to look at the results of the *Breslow-Day test* for homogeneity of odds ratios given in Display 3.13. Here there is no evidence against homogeneity. The Breslow-Day test is described in Agresti [1996]).

```
The FREQ Procedure

                  Summary Statistics for level by bronch
                          Controlling for agegrp

         Cochran-Mantel-Haenszel Statistics (Based on Table Scores)

         Statistic   Alternative Hypothesis     DF    Value      Prob
         ------------------------------------------------------------
             1       Nonzero Correlation         1    0.2215    0.6379
             2       Row Mean Scores Differ      1    0.2215    0.6379
             3       General Association         1    0.2215    0.6379

            Estimates of the Common Relative Risk (Row1/Row2)

  Type of Study    Method            Value    95% Confidence    Limits
  --------------------------------------------------------------------
  Case-Control     Mantel-Haenszel   1.1355        0.6693        1.9266
  (Odds Ratio)     Logit             1.1341        0.6678        1.9260

  Cohort           Mantel-Haenszel   1.1291        0.6808        1.8728
  (Col1 Risk)      Logit             1.1272        0.6794        1.8704

  Cohort I         Mantel-Haenszel   0.9945        0.9725        1.0170
  (Col2 Risk)      Logit             0.9945        0.9729        1.0166
```

```
                 Breslow-Day Test for
             Homogeneity of the Odds Ratios
           ------------------------------------
           Chi-Square                    0.1173
           DF                                 2
           Pr > ChiSq                    0.9430

               Total Sample Size = 1462
```

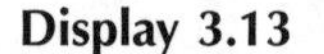

Display 3.13

Exercises

3.1 For the oral contraceptives data, construct a confidence interval for the difference in the proportion of women suffering blood clots who used oral contraceptives and the corresponding proportion for women not suffering blood clots.

3.2 For the piston ring data, the "residuals" used in the text were simply observed frequency minus expected under independence. Those are not satisfactory for a number of reasons, as discussed in Everitt (1992). More suitable residuals are r and r_{adj} given by:

$$r = \frac{\text{Observed} - \text{Expected}}{\sqrt{\text{Expected}}} \quad \text{and}$$

$$r_{adj} = \frac{r}{\sqrt{\left(1 - \frac{\text{Row total}}{\text{Sample size}}\right) - \left(1 - \frac{\text{Column total}}{\text{Sample size}}\right)}}$$

Calculate both for the piston ring data and compare what each of the three types have to say about the data.

3.3 In the data given in Display 3.5, the frequencies for the Keral and Andhra regions are identical. Reanalyse the data after simply summing the frequencies for those two regions and reducing the number of columns of the table by one.

Chapter 4

Multiple Regression: Determinants of Crime Rate in the United States

4.1 Description of Data

The data set of interest in this chapter is shown in Display 4.1 and consists of crime rates for 47 states in the United States, along with the values of 13 explanatory variables possibly associated with crime. (The data were originally given in Vandaele [1978] and also appear in Table 134 of *SDS*.)

A full description of the 14 variables in these data is as follows:

R	Crime rate: the number of offences known to the police per 1,000,000 population
Age	Age distribution: the number of males aged 14 to 24 years per 1000 of total state population
S	Binary variable distinguishing southern states (S = 1) from the rest
Ed	Educational level: mean number of years of schooling × 10 of the population 25 years old and over
Ex0	Police expenditure: per capita expenditure on police protection by state and local governments in 1960
Ex1	Police expenditure: as Ex0, but for 1959

LF	Labour force participation rate per 1000 civilian urban males in the age group 14 to 24 years
M	Number of males per 1000 females
N	State population size in hundred thousands
NW	Number of non-whites per 1000
U1	Unemployment rate of urban males per 1000 in the age group 14 to 24 years
U2	Unemployment rate of urban males per 1000 in the age group 35 to 39 years
W	Wealth, as measured by the median value of transferable goods and assets or family income (unit 10 dollars)
X	Income inequality: the number of families per 1000 earning below one half of the median income

The main question of interest about these data concerns how the crime rate depends on the other variables listed. The central method of analysis will be *multiple regression*.

R	*Age*	*S*	*Ed*	*Ex0*	*Ex1*	*LF*	*M*	*N*	*NW*	*U1*	*U2*	*W*	*X*
79.1	151	1	91	58	56	510	950	33	301	108	41	394	261
163.5	143	0	113	103	95	583	1012	13	102	96	36	557	194
57.8	142	1	89	45	44	533	969	18	219	94	33	318	250
196.9	136	0	121	149	141	577	994	157	80	102	39	673	167
123.4	141	0	121	109	101	591	985	18	30	91	20	578	174
68.2	121	0	110	118	115	547	964	25	44	84	29	689	126
96.3	127	1	111	82	79	519	982	4	139	97	38	620	168
155.5	131	1	109	115	109	542	969	50	179	79	35	472	206
85.6	157	1	90	65	62	553	955	39	286	81	28	421	239
70.5	140	0	118	71	68	632	1029	7	15	100	24	526	174
167.4	124	0	105	121	116	580	966	101	106	77	35	657	170
84.9	134	0	108	75	71	595	972	47	59	83	31	580	172
51.1	128	0	113	67	60	624	972	28	10	77	25	507	206
66.4	135	0	117	62	61	595	986	22	46	77	27	529	190
79.8	152	1	87	57	53	530	986	30	72	92	43	405	264
94.3	142	1	88	81	77	497	956	33	321	116	47	427	157
53.9	143	0	110	66	63	537	977	10	6	114	35	487	166
92.9	135	1	104	123	115	537	978	31	170	89	34	631	165
75.0	130	0	116	128	128	536	934	51	24	78	34	627	135
122.5	125	0	108	113	105	567	985	78	94	130	58	626	166
74.2	126	0	108	74	67	602	984	34	12	102	33	557	195
43.9	157	1	89	47	44	512	962	22	423	97	34	288	276
121.6	132	0	96	87	83	564	953	43	92	83	32	513	227

96.8	131	0	116	78	73	574	1038	7	36	142	42	540	179
52.3	130	0	116	63	57	641	984	14	26	70	21	486	196
199.3	131	0	121	160	143	631	1071	3	77	102	41	674	152
34.2	135	0	109	69	71	540	965	6	4	80	22	564	139
121.6	152	0	112	82	76	571	1018	10	79	103	28	537	215
104.3	119	0	107	166	157	521	928	168	89	92	36	637	154
69.6	166	1	89	58	54	521	973	46	2554	72	26	396	237
37.3	140	0	93	55	54	535	1045	6	20	135	40	453	200
75.4	125	0	109	90	81	586	964	97	82	105	43	617	163
107.2	147	1	104	63	64	560	972	23	95	76	24	462	233
92.3	126	0	118	97	97	542	990	18	21	102	35	589	166
65.3	123	0	102	97	87	526	958	113	76	124	50	572	158
127.2	150	0	100	109	98	531	964	9	24	87	38	559	153
83.1	177	1	87	58	56	638	974	24	349	76	28	382	254
56.6	133	0	104	51	47	599	1024	7	40	99	27	425	225
82.6	149	1	88	61	54	515	953	36	165	86	35	395	251
115.1	145	1	104	82	74	560	981	96	126	88	31	488	228
88.0	148	0	122	72	66	601	998	9	19	84	20	590	144
54.2	141	0	109	56	54	523	968	4	2	107	37	489	170
82.3	162	1	99	74	70	522	996	40	208	73	27	496	221
103.0	136	0	121	95	96	574	1012	29	36	111	37	622	162
45.5	139	1	88	46	41	480	9968	19	49	135	53	457	249
50.8	126	0	104	106	97	599	989	40	24	78	25	593	171
84.9	130	0	121	90	91	623	1049	3	22	113	40	588	160

Display 4.1

4.2 The Multiple Regression Model

The multiple regression model has the general form:

$$y_i = \beta_0 + \beta_1 x_{1i} + \beta_2 x_{2i} + \cdots \beta_p x_{pi} + \epsilon_i \quad (4.1)$$

where y_i is the value of a continuous response variable for observation i, and x_{1i}, x_{2i}, $\cdots$, x_{pi} are the values of p explanatory variables for the same observation. The term ϵ_i is the residual or error for individual i and represents the deviation of the observed value of the response for this individual from that expected by the model. The regression coefficients, β_0, β_1, $\cdots$, β_p, are generally estimated by least-squares.

Significance tests for the regression coefficients can be derived by assuming that the residual terms are normally distributed with zero mean and constant variance σ^2. The estimated regression coefficient corresponding to a particular explanatory variable gives the change in the response variable associated with a unit change in the explanatory variable, conditional on all other explanatory variables remaining constant.

For n observations of the response and explanatory variables, the regression model can be written concisely as:

$$\boldsymbol{y} = \boldsymbol{X\beta} + \in \tag{4.2}$$

where $\boldsymbol{y}$ is the $n \times 1$ vector of responses; $\boldsymbol{X}$ is an $n \times (p + 1)$ matrix of known constraints, the first column containing a series of ones corresponding to the term β_0 in Eq. (4.1); and the remaining columns containing values of the explanatory variables. The elements of the vector $\boldsymbol{\beta}$ are the regression coefficients $\beta_0, \beta_1, \cdots, \beta_p$, and those of the vector $\boldsymbol{\in}$, the residual terms $\in_1, \in_2, \cdots, \in_n$.

The regression coefficients can be estimated by least-squares, resulting in the following estimator for $\boldsymbol{\beta}$:

$$\hat{\boldsymbol{\beta}} = (\boldsymbol{X'X})^{-1} \boldsymbol{X'y} \tag{4.3}$$

The variances and covariances of the resulting estimates can be found from

$$\boldsymbol{S}_{\hat{\boldsymbol{\beta}}} = s^2(\boldsymbol{X'X})^{-1} \tag{4.4}$$

where s^2 is defined below.

The variation in the response variable can be partitioned into a part due to regression on the explanatory variables and a residual. These can be arranged in an analysis of variance table as follows:

Source	*DF*	*SS*	*MS*	*F*
Regression	p	RGSS	RGSS/p	RGMS/RSMS
Residual	$n-p-1$	RSS	RSS/$(n-p-1)$	

Note: DF: degrees of freedom, *SS:* sum of squares, *MS*: mean square.

The residual mean square s^2 gives an estimate of σ^2, and the F-statistic is a test that $\beta_1, \beta_2, \cdots, \beta_p$ are all zero.

A measure of the fit of the model is provided by the *multiple correlation coefficient*, R, defined as the correlation between the observed values of the response variable and the values predicted by the model; that is

$$\hat{y}_i = \hat{\beta}_0 + \hat{\beta}_1 x_{i1} + \cdots \hat{\beta}_p x_{ip} \quad (4.5)$$

The value of R^2 gives the proportion of the variability of the response variable accounted for by the explanatory variables.

For complete details of multiple regression, see, for example, Rawlings (1988).

4.3 Analysis Using SAS

Assuming that the data are available as an ASCII file **uscrime.dat**, they can be read into SAS for analysis using the following instructions:

```
data uscrime;
   infile 'uscrime.dat' expandtabs;
   input R Age S Ed Ex0 Ex1 LF M N NW U1 U2 W X;
run;
```

Before undertaking a formal regression analysis of these data, it may be helpful to examine them graphically using a *scatterplot matrix*. This is essentially a grid of scatterplots for each pair of variables. Such a display is often useful in assessing the general relationships between the variables, in identifying possible outliers, and in highlighting potential *multicollinearity* problems amongst the explanatory variables (i.e., one explanatory variable being essentially predictable from the remainder). Although this is only available routinely in the SAS/INSIGHT module, for those with access to SAS/IML, we include a macro, listed in Appendix A, which can be used to produce a scatterplot matrix. The macro is invoked as follows:

```
%include 'scattmat.sas';
%scattmat(uscrime,R--X);
```

This assumes that the macro is stored in the file **'scattmat.sas'** in the current directory. Otherwise, the full pathname is needed. The macro is called with the **%scattmat** statement and two parameters are passed: the name of the SAS data set and the list of variables to be included in the scatterplot matrix. The result is shown in Display 4.2.

The individual relationships of crime rate to each of the explanatory variables shown in the first column of this plot do not appear to be

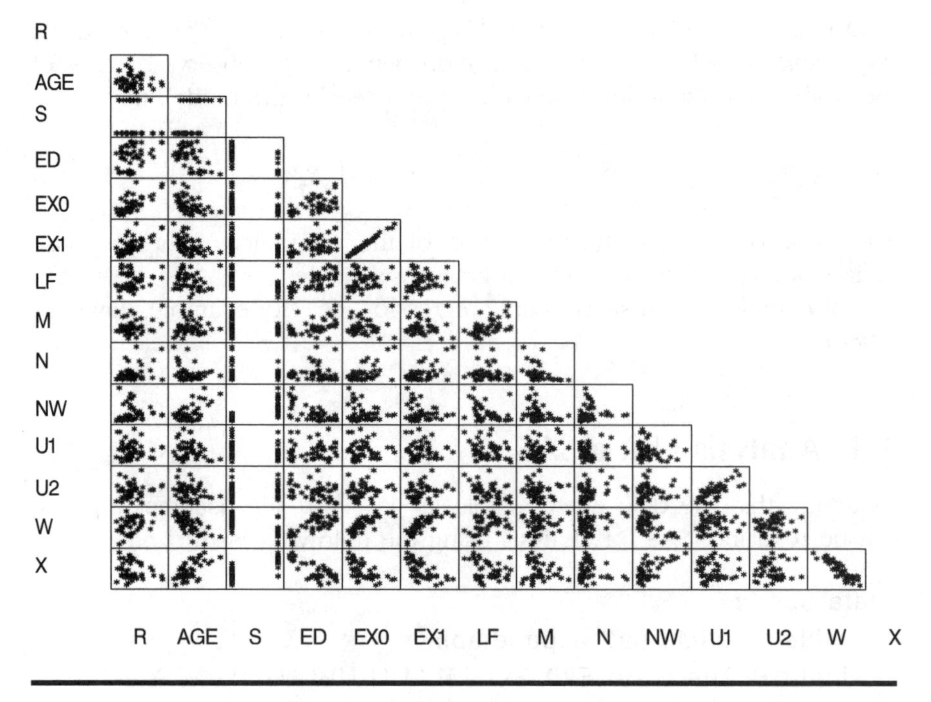

Display 4.2

particularly strong, apart perhaps from Ex0 and Ex1. The scatterplot matrix also clearly highlights the very strong relationship between these two variables. Highly correlated explanatory variables, multicollinearity, can cause several problems when applying the multiple regression model, including:

1. It severely limits the size of the multiple correlation coefficient R because the explanatory variables are primarily attempting to explain much of the same variability in the response variable (see Dizney and Gromen [1967] for an example).
2. It makes determining the importance of a given explanatory variable (see later) difficult because the effects of explanatory variables are confounded due to their intercorrelations.
3. It increases the variances of the regression coefficients, making use of the predicted model for prediction less stable. The parameter estimates become unreliable.

Spotting multicollinearity amongst a set of explanatory variables might not be easy. The obvious course of action is to simply examine the correlations between these variables, but whilst this *is* often helpful, it is by no means foolproof — more subtle forms of multicollinearity may be missed. An alternative and generally far more useful approach is to examine what are known as the *variance inflation factors* of the explanatory variables. The variance inflation factor VIF_j for the *j*th variable is given by

$$VIF_j = \frac{1}{1 - R_j^2} \tag{4.6}$$

where R_j^2 is the square of the multiple correlation coefficient from the regression of the *j*th explanatory variable on the remaining explanatory variables. The variance inflation factor of an explanatory variable indicates the strength of the linear relationship between the variable and the remaining explanatory variables. A rough rule of thumb is that variance inflation factors greater than 10 give some cause for concern.

How can multicollinearity be combatted? One way is to combine in some way explanatory variables that are highly correlated. An alternative is simply to select one of the set of correlated variables. Two more complex possibilities are *regression on principal components* and *ridge regression*, both of which are described in Chatterjee and Price (1991).

The analysis of the crime rate data begins by looking at the variance inflation factors of the 13 explanatory variables, obtained using the following SAS instructions:

```
proc reg data=uscrime;
   model R= Age--X / vif;
run;
```

The vif option in the model statement requests that variance inflation factors be included in the output shown in Display 4.3.

The REG Procedure
Model: MODEL1
Dependent Variable: R

Analysis of Variance

Source	DF	Sum of Squares	Mean Square	F Value	Pr > F
Model	13	52931	4071.58276	8.46	<.0001
Error	33	15879	481.17275		
Corrected Total	46	68809			

Root MSE	21.93565	R-Square	0.7692
Dependent Mean	90.50851	Adj R-Sq	0.6783
Coeff Var	24.23601		

Parameter Estimates

Variable	DF	Parameter Estimate	Standard Error	t Value	Pr > |t|	Variance Inflation
Intercept	1	-691.83759	155.88792	-4.44	<.0001	0
Age	1	1.03981	0.42271	2.46	0.0193	2.69802
S	1	-8.30831	14.91159	-0.56	0.5812	4.87675
Ed	1	1.80160	0.64965	2.77	0.0091	5.04944
Ex0	1	1.60782	1.05867	1.52	0.1384	94.63312
Ex1	1	-0.66726	1.14877	-0.58	0.5653	98.63723
LF	1	-0.04103	0.15348	-0.27	0.7909	3.67756
M	1	0.16479	0.20993	0.78	0.4381	3.65844
N	1	-0.04128	0.12952	-0.32	0.7520	2.32433
NW	1	0.00717	0.06387	0.11	0.9112	4.12327
U1	1	-0.60168	0.43715	-1.38	0.1780	5.93826
U2	1	1.79226	0.85611	2.09	0.0441	4.99762
W	1	0.13736	0.10583	1.30	0.2033	9.96896
X	1	0.79293	0.23509	3.37	0.0019	8.40945

Display 4.3

Concentrating for now on the variance inflation factors in Display 4.3, we see that those for **Ex0** and **Ex1** are well above the value 10. As a consequence, we simply drop variable **Ex0** from consideration and now regress crime rate on the remaining 12 explanatory variables using the following:

```
proc reg data=uscrime;
  model R= Age--Ed Ex1--X / vif;
run;
```

The output is shown in Display 4.4. The square of the multiple correlation coefficient is 0.75, indicating that the 12 explanatory variables account for 75% of the variability in the crime rates of the 47 states. The variance inflation factors are now all less than 10.

The REG Procedure

Model: MODEL1

Dependent Variable: R

Analysis of Variance

Source	DF	Sum of Squares	Mean Square	F Value	Pr > F
Model	12	51821	4318.39553	8.64	<.0001
Error	34	16989	499.66265		
Corrected Total	46	68809			

Root MSE	22.35314	R-Square	0.7531
Dependent Mean	90.50851	Adj R-Sq	0.6660
Coeff Var	24.69727		

Parameter Estimates

Variable	DF	Parameter Estimate	Standard Error	t Value	Pr > |t|	Variance Inflation
Intercept	1	-739.89065	155.54826	-4.76	<.0001	0
Age	1	1.08541	0.42967	2.53	0.0164	2.68441
S	1	-8.16412	15.19508	-0.54	0.5946	4.87655
Ed	1	1.62669	0.65153	2.50	0.0175	4.89076
Ex1	1	1.02965	0.27202	3.79	0.0006	5.32594
LF	1	0.00509	0.15331	0.03	0.9737	3.53357
M	1	0.18686	0.21341	0.88	0.3874	3.64093
N	1	-0.01639	0.13092	-0.13	0.9011	2.28711
NW	1	-0.00213	0.06478	-0.03	0.9740	4.08533
U1	1	-0.61879	0.44533	-1.39	0.1737	5.93432
U2	1	1.94296	0.86653	2.24	0.0316	4.93048
W	1	0.14739	0.10763	1.37	0.1799	9.93013
X	1	0.81550	0.23908	3.41	0.0017	8.37584

Display 4.4

The adjusted R^2 statistic given in Display 4.4 is the square of the multiple correlation coefficient adjusted for the number of parameters in the model. The statistic is calculated as:

$$\text{adj}R^2 = 1 - \frac{(n-1)(1-R^2)}{n-p} \tag{4.7}$$

where n is the number of observations used in fitting the model and i is an indicator variable that is 1 if the model includes an intercept and 0 otherwise.

The main features of interest in Display 4.4 are the analysis of variance table and the parameter estimates. In the former, the F-test is for the hypothesis that *all* the regression coefficients in the regression equation are zero. Here, the evidence against this hypothesis is very strong (the relevant P-value is 0.0001). In general, however, this overall test is of little real interest because it is most unlikely in general that *none* of the explanatory variables will be related to the response. The more relevant question is whether a subset of the regression coefficients is zero, implying that not *all* the explanatory variables are informative in determining the response. It might be thought that the nonessential variables can be identified by simply examining the estimated regression coefficients and their standard errors as given in Display 4.4, with those regression coefficients significantly different from zero identifying the explanatory variables needed in the derived regression equation, and those not different from zero corresponding to variables that can be omitted. Unfortunately, this very straightforward approach is not in general suitable, simply because the explanatory variables are correlated in most cases. Consequently, removing a particular explanatory variable from the regression will alter the estimated regression coefficients (and their standard errors) of the remaining variables. The parameter estimates and their standard errors are *conditional* on the other variables in the model. A more involved procedure is thus necessary for identifying subsets of the explanatory variables most associated with crime rate. A number of methods are available, including:

- *Forward selection.* This method starts with a model containing *none* of the explanatory variables and then considers variables one by one for inclusion. At each step, the variable added is one that results in the biggest increase in the *regression sum of squares.* An F-type statistic is used to judge when further additions would not represent a significant improvement in the model.
- *Backward elimination.* This method starts with a model containing *all* the explanatory variables and eliminates variables one by one, at each stage choosing the variable for exclusion as the one leading to the smallest decrease in the regression sum of squares. Once again, an F-type statistic is used to judge when further exclusions would represent a significant deterioration in the model.

- *Stepwise regression.* This method is, essentially, a combination of forward selection and backward elimination. Starting with no variables in the model, variables are added as with the forward selection method. Here, however, with each addition of a variable, a backward elimination process is considered to assess whether variables entered earlier might now be removed because they no longer contribute significantly to the model.

In the best of all possible worlds, the final model selected by each of these procedures would be the same. This is often the case, but it is in no way guaranteed. It should also be stressed that none of the automatic procedures for selecting subsets of variables are foolproof. They must be used with care, and warnings such as the following given in Agresti (1996) must be noted:

> Computerized variable selection procedures should be used with caution. When one considers a large number of terms for potential inclusion in a model, one or two of them that are not really important may look impressive simply due to chance. For instance, when all the true effects are weak, the largest sample effect may substantially overestimate its true effect. In addition it often makes sense to include certain variables of special interest in a model and report their estimated effects even if they are not statistically significant at some level.

In addition, the comments given in McKay and Campbell (1982a;b) concerning the validity of the *F*-tests used to judge whether variables should be included or eliminated, should be considered.

Here, we apply a stepwise procedure using the following SAS code:

```
proc reg data=uscrime;
   model R= Age--Ed Ex1--X / selection=stepwise sle=.05
sls=.05;
   plot student.*(ex1 x ed age u2);
   plot student.*predicted. cookd.*obs.;
   plot npp.*residual.;
run;
```

The proc, model, and run statements specify the regression analysis and produce the output shown in Display 4.5. The significance levels required for variables to enter and stay in the regression are specified with the sle and sls options, respectively. The default for both is $P = 0.15$. (The plot statements in this code are explained later.)

Display 4.5 shows the variables entered at each stage in the variable selection procedure. At step one, variable Ex1 is entered. This variable is the best single predictor of the crime rate. The square of the multiple correlation coefficient is observed to be 0.4445. The variable Ex1 explains 44% of the variation in crime rates. The analysis of variance table shows both the regression and residual or error sums of squares. The *F*-statistics is highly significant, confirming the strong relationship between crime rate and Ex1. The estimated regression coefficient is 0.92220, with a standard error of 0.15368. This implies that a unit increase in Ex1 is associated with an estimated increase in crime rate of 0.92. This appears strange but perhaps police expenditures increase as the crime rate increases.

At step two, variable X is entered. The R-square value increases to 0.5550. The estimated regression coefficient of X is 0.42312, with a standard error of 0.12803. In the context of regression, the type II sums of squares and *F*-tests based on them are equivalent to type III sums of squares described in Chapter 6.

In this application of the stepwise option, the default significance levels for the *F*-tests used to judge entry of a variable into an existing model and to judge removal of a variable from a model are each set to 0.05. With these values, the stepwise procedure eventually identifies a subset of five explanatory variables as being important in the prediction of the crime rate. The final results are summarised at the end of Display 4.5. The selected five variables account for just over 70% of the variation in crime rates compared to the 75% found when using 12 explanatory variables in the previous analysis. (Notice that in this example, the stepwise procedure gives the same results as would have arisen from using forward selection with the same entry criterion value of 0.05 because none of the variables entered in the "forward" phase are ever removed.)

The statistic C_p was suggested by Mallows (1973) as a possible alternative criterion useful for selecting informative subsets of variables. It is defined as:

$$C_p = \frac{SSE_p}{s^2} - (n - 2p) \qquad (4.8)$$

where s^2 is the mean square error from the regression, including all the explanatory variables available; and SSE_p is the error sum of squares for a model that includes just a subset of the explanatory variable. If C_p is plotted against p, Mallows recommends accepting the model where C_p first approaches p (see Exercise 4.2).

(The Bounds on condition number given in Display 4.5 are fully explained in Berk [1977]. Briefly, the condition number is the ratio of the largest and smallest eigenvalues of a matrix and is used as a measure of the numerical stability of the matrix. Very large values are indicative of possible numerical problems.)

The REG Procedure
Model: MODEL1
Dependent Variable: R

Stepwise Selection: Step 1

Variable Ex1 Entered: R-Square = 0.4445 and C(p) = 33.4977

Analysis of Variance

Source	DF	Sum of Squares	Mean Square	F Value	Pr > F
Model	1	30586	30586	36.01	<.0001
Error	45	38223	849.40045		
Corrected Total	46	68809			

Variable	Parameter Estimate	Standard Error	Type II SS	F Value	Pr > F
Intercept	16.51642	13.04270	1362.10230	1.60	0.2119
Ex1	0.92220	0.15368	30586	36.01	<.0001

Bounds on condition number: 1, 1

Stepwise Selection: Step 2

Variable X Entered: R-Square = 0.5550 and C(p) = 20.2841

Analysis of Variance

Source	DF	Sum of Squares	Mean Square	F Value	Pr > F
Model	2	38188	19094	27.44	<.0001
Error	44	30621	695.94053		
Corrected Total	46	68809			

Variable	Parameter Estimate	Standard Error	Type II SS	F Value	Pr > F
Intercept	-96.96590	36.30976	4963.21825	7.13	0.0106
Ex1	1.31351	0.18267	35983	51.70	<.0001
X	0.42312	0.12803	7601.63672	10.92	0.0019

Bounds on condition number: 1.7244, 6.8978

Stepwise Selection: Step 3

Variable Ed Entered: R-Square = 0.6378 and C(p) = 10.8787

Analysis of Variance

Source	DF	Sum of Squares	Mean Square	F Value	Pr > F
Model	3	43887	14629	25.24	<.0001
Error	43	24923	579.59373		
Corrected Total	46	68809			

The REG Procedure
Model: MODEL1
Dependent Variable: R

Stepwise Selection: Step 3

Variable	Parameter Estimate	Standard Error	Type II SS	F Value	Pr > F
Intercept	-326.10135	80.23552	9574.04695	16.52	0.0002
Ed	1.55544	0.49605	5698.85308	9.83	0.0031
Ex1	1.31222	0.16671	35912	61.96	<.0001
X	0.75779	0.15825	13291	22.93	<.0001

Bounds on condition number: 3.1634, 21.996

Stepwise Selection: Step 4

Variable Age Entered: R-Square = 0.6703 and C(p) = 8.4001

Analysis of Variance

Source	DF	Sum of Squares	Mean Square	F Value	Pr > F
Model	4	46125	11531	21.35	<.0001
Error	42	22685	540.11277		
Corrected Total	46	68809			

Variable	Parameter Estimate	Standard Error	Type II SS	F Value	Pr > F
Intercept	-420.16714	90.19340	11721	21.70	<.0001
Age	0.73451	0.36085	2237.79373	4.14	0.0481
Ed	1.63349	0.48039	6245.05569	11.56	0.0015
Ex1	1.36844	0.16328	37937	70.24	<.0001
X	0.65225	0.16132	8829.50458	16.35	0.0002

Bounds on condition number: 3.5278, 38.058

Stepwise Selection: Step 5

Variable U2 Entered: R-Square = 0.7049 and C(p) = 5.6452

Analysis of Variance

Source	DF	Sum of Squares	Mean Square	F Value	Pr > F
Model	5	48500	9700.08117	19.58	<.0001
Error	41	20309	495.33831		
Corrected Total	46	68809			

Variable	Parameter Estimate	Standard Error	Type II SS	F Value	Pr > F
Intercept	-528.85572	99.61621	13961	28.18	<.0001
Age	1.01840	0.36909	3771.26606	7.61	0.0086
Ed	2.03634	0.49545	8367.48486	16.89	0.0002
Ex1	1.29735	0.15970	32689	65.99	<.0001
U2	0.99014	0.45210	2375.86580	4.80	0.0343
X	0.64633	0.15451	8667.44486	17.50	0.0001

The REG Procedure
Model: MODEL1
Dependent Variable: R

Stepwise Selection: Step 5

Bounds on condition number: 3.5289, 57.928

All variables left in the model are significant at the 0.0500 level.

No other variable met the 0.0500 significance level for entry into the model.

Summary of Stepwise Selection

Step	Variable Entered	Variable Removed	Number Vars In	Partial R-Square	Model R-Square	C(p)	F Value	Pr > F
1	Ex1		1	0.4445	0.4445	33.4977	36.01	<.0001
2	X		2	0.1105	0.5550	20.2841	10.92	0.0019
3	Ed		3	0.0828	0.6378	10.8787	9.83	0.0031
4	Age		4	0.0325	0.6703	8.4001	4.14	0.0481
5	U2		5	0.0345	0.7049	5.6452	4.80	0.0343

Display 4.5

Having arrived at a final multiple regression model for a data set, it is important to go further and check the assumptions made in the modelling process. Most useful at this stage is an examination of *residuals* from the fitted model, along with many other regression diagnostics now available. Residuals at their simplest are the difference between the observed and fitted values of the response variable — in our example, crime rate. The most useful ways of examining the residuals are graphical, and the most useful plots are

- A plot of the residuals against each explanatory variable in the model; the presence of a curvilinear relationship, for example, would suggest that a higher-order term (e.g., a quadratic) in the explanatory variable is needed in the model.
- A plot of the residuals against predicted values of the response variable; if the variance of the response appears to increase with the predicted value, a transformation of the response may be in order.
- A normal probability plot of the residuals; after all systematic variation has been removed from the data, the residuals should look like a sample from the normal distribution. A plot of the ordered residuals against the expected order statistics from a normal distribution provides a graphical check of this assumption.

Unfortunately, the simple observed-fitted residuals have a distribution that is scale dependent (see Cook and Weisberg [1982]), which makes them less helpful than they might be. The problem can be overcome, however, by using *standardised* or *studentised* residuals (both are explicitly defined in Cook and Weisberg [1982]) .

A variety of other diagnostics for regression models have been developed in the past decade or so. One that is often used is the *Cook's distance statistic* (Cook [1977; 1979]). This statistic can be obtained for each of the *n* observations and measures the change to the estimates of the regression coefficients that would result from deleting the particular observation. It can be used to identify any observations having an undue influence of the estimation and fitting process.

Plots of residuals and other diagnostics can be found using the plot statement to produce high-resolution diagnostic plots. Variables mentioned in the model or var statements can be plotted along with diagnostic statistics. The latter are represented by keywords that end in a period. The first plot statement produces plots of the studentised residual against the five predictor variables. The results are shown in Display 4.6 through Display 4.10. The next plot statement produces a plot of the studentised residuals against the predicted values and an index plot of Cook's distance statistic. The resulting plots are shown in Displays 4.11 and 4.12. The final plot statement specifies a normal probability plot of the residuals, which is shown in Display 4.13.

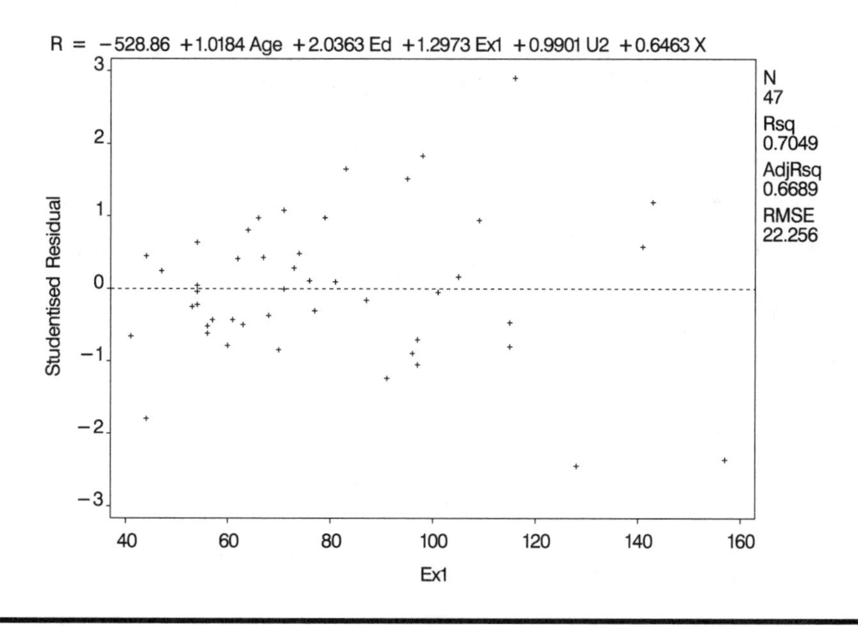

Display 4.6

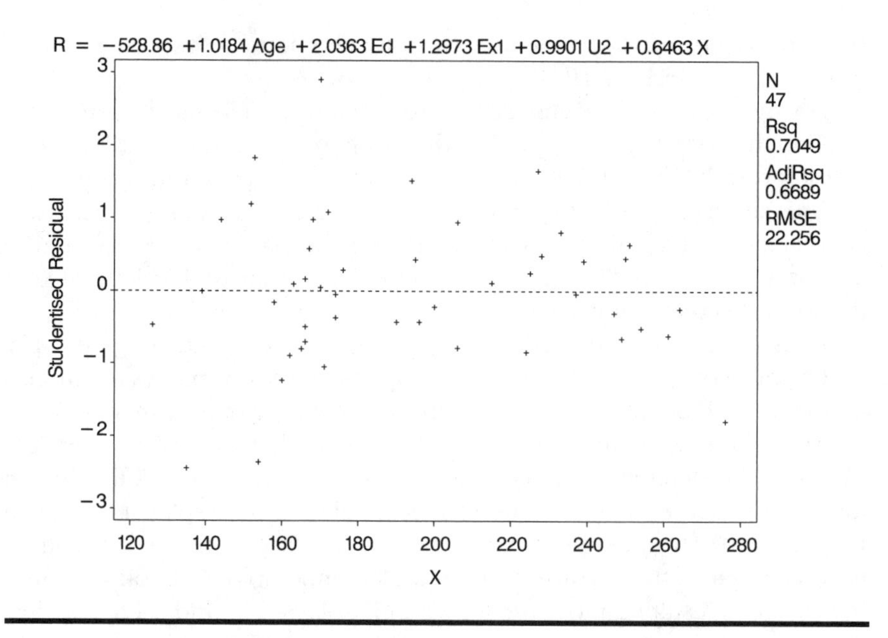

Display 4.7

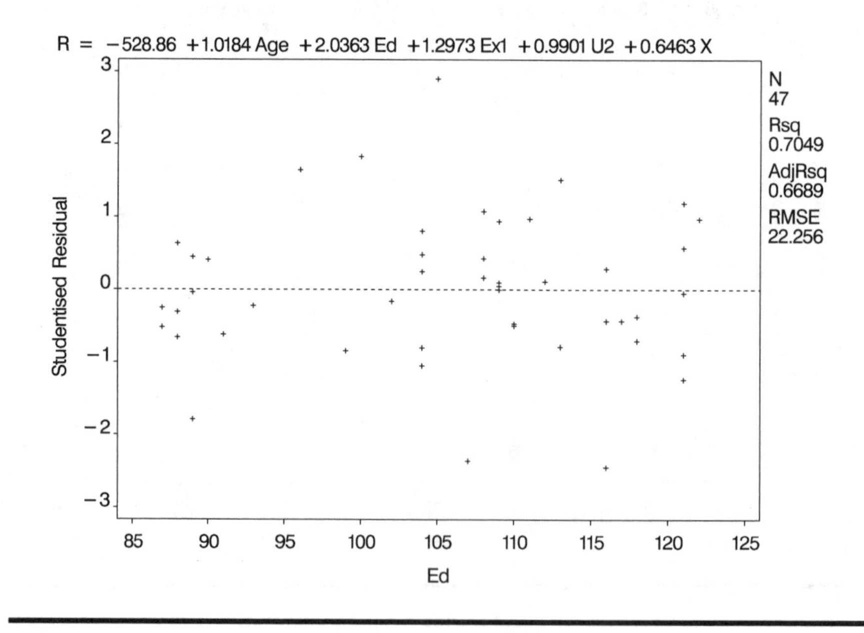

Display 4.8

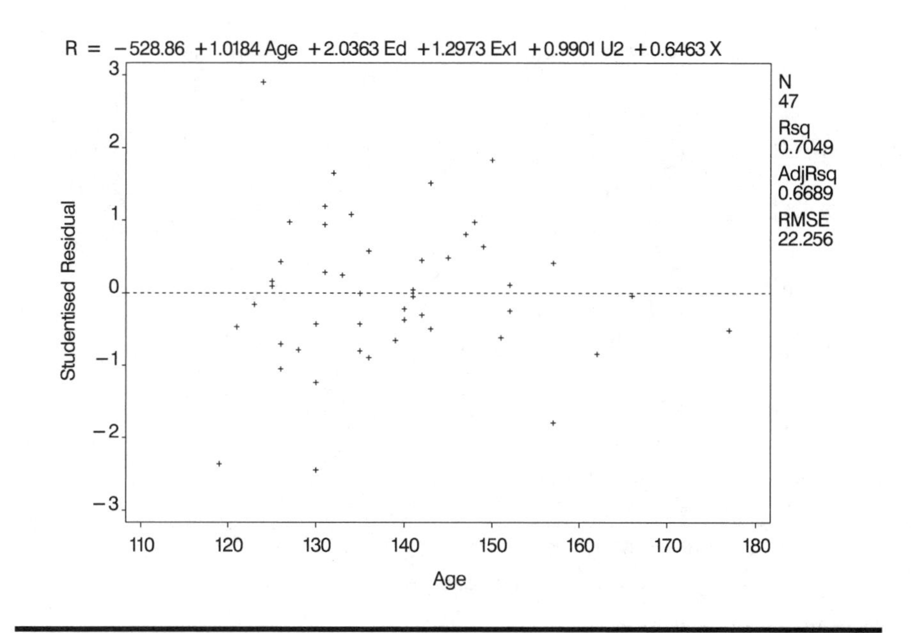

Display 4.9

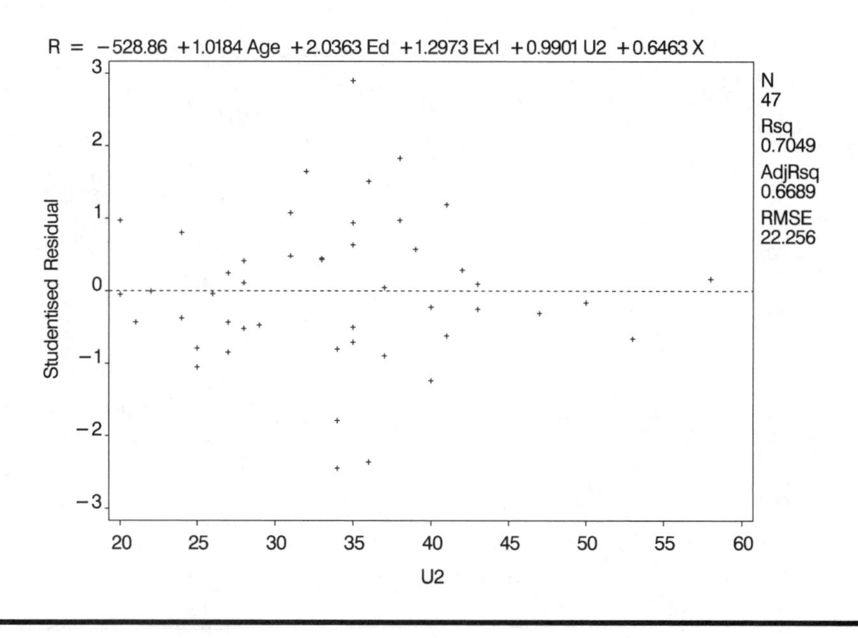

Display 4.10

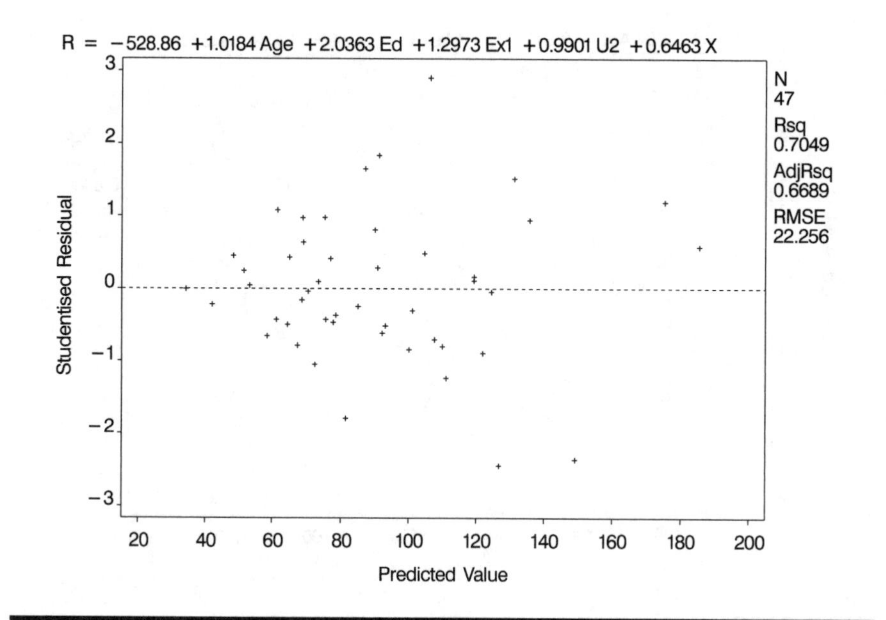

Display 4.11

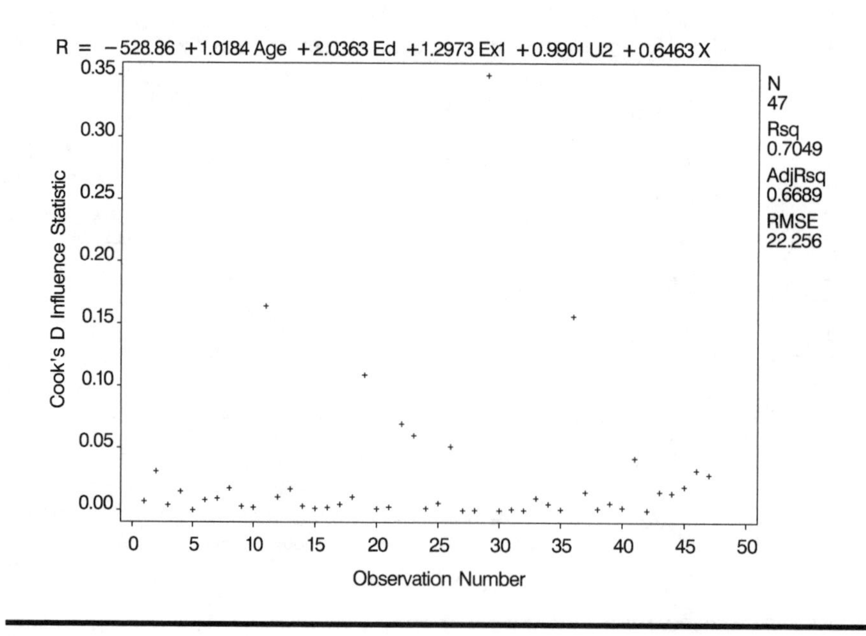

Display 4.12

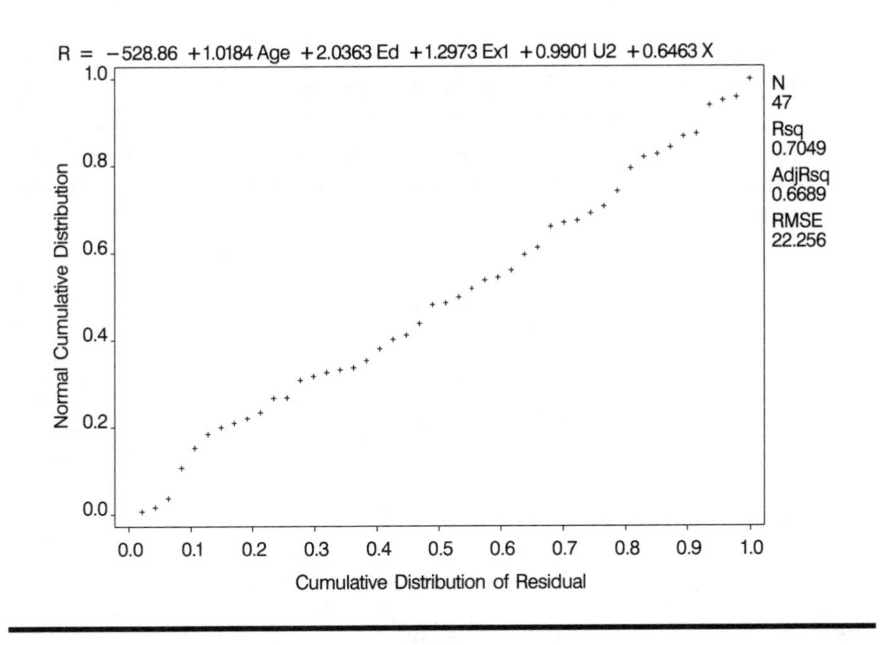

Display 4.13

Display 4.6 suggests increasing variability of the residuals with increasing values of Ex1. And Display 4.13 indicates a number of relatively large values for the Cook's distance statistic although there are no values greater than 1, which is the usually accepted threshold for concluding that the corresponding observation has undue influence on the estimated regression coefficients.

Exercises

4.1 Find the subset of five variables considered by the C_p option to be optimal. How does this subset compare with that chosen by the stepwise option?

4.2 Apply the C_p criterion to exploring all possible subsets of the five variables chosen by the stepwise procedure (see Display 4.5). Produce a plot of the number of variables in a subset against the corresponding value of C_p.

4.3 Examine some of the other regression diagnostics available with proc reg on the U.S. crime rate data.

4.4 In the text, the problem of the high variance inflation factors associated with variables Ex0 and Ex1 was dealt with by excluding

ExO. An alternative is to use the average of the two variables as an explanatory variable. Investigate this possibility.

4.5 Investigate the regression of crime rate on the two variables *Age* and *S*. Consider the possibility of an interaction of the two variables in the regression model, and construct some plots that illustrate the models fitted.

Chapter 5

Analysis of Variance I: Treating Hypertension

5.1 Description of Data

Maxwell and Delaney (1990) describe a study in which the effects of three possible treatments for hypertension were investigated. The details of the treatments are as follows:

Treatment	*Description*	*Levels*
Drug	Medication	Drug X, drug Y, drug Z
Biofeed	Psychological feedback	Present, absent
Diet	Special diet	Present, absent

All 12 combinations of the three treatments were included in a $3 \times 2 \times 2$ design. Seventy-two subjects suffering from hypertension were recruited to the study, with six being randomly allocated to each of 12 treatment combinations. Blood pressure measurements were made on each subject after treatment, leading to the data in Display 5.1.

Treatment		Special Diet											
Biofeedback	Drug	No						Yes					
Present	X	170	175	165	180	160	158	161	173	157	152	181	190
	Y	186	194	201	215	219	209	164	166	159	182	187	174
	Z	180	187	199	170	204	194	162	184	183	156	180	173
Absent	X	173	194	197	190	176	198	164	190	169	164	176	175
	Y	189	194	217	206	199	195	171	173	196	199	180	203
	Z	202	228	190	206	224	204	205	199	170	160	179	179

Display 5.1

Questions of interest concern differences in mean blood pressure for the different levels of the three treatments and the possibility of interactions between the treatments.

5.2 Analysis of Variance Model

A possible model for these data is

$$y_{ijkl} = \mu + \alpha_i + \beta_j + \gamma_k + (\alpha\beta)_{ij} + (\alpha\gamma)_{ik} + (\beta\gamma)_{jk} + (\alpha\beta\gamma)_{ijk} + \epsilon_{ijkl} \quad (5.1)$$

where y_{ijkl} represents the blood pressure of the lth subject for the ith drug, the jth level of biofeedback, and the kth level of diet; μ is the overall mean; α_i, β_j, and γ_k are the main effects of drugs, biofeedback, and diets; $(\alpha\beta)_{ij}$, $(\alpha\gamma)_{ik}$, and $(\beta\gamma)_{jk}$ are the first-order interaction terms between pairs of treatments, $(\alpha\beta\gamma)_{ijk}$ represents the second-order interaction term of the three treatments; and ϵ_{ijkl} represents the residual or error terms assumed to be normally distributed with zero mean and variance σ^2. (The model as specified is over-parameterized and the parameters have to be constrained in some way, commonly by requiring them to sum to zero or setting one parameter at zero; see Everitt [2001] for details.)

Such a model leads to a partition of the variation in the observations into parts due to main effects, first-order interactions between pairs of factors, and a second-order interaction between all three factors. This partition leads to a series of F-tests for assessing the significance or otherwise of these various components. The assumptions underlying these F-tests include:

- The observations are independent of one another.
- The observations in each cell arise from a population having a normal distribution.
- The observations in each cell are from populations having the same variance.

5.3 Analysis Using SAS

It is assumed that the 72 blood pressure readings shown in Display 5.1 are in the ASCII file hypertension.dat. The SAS code used for reading and labelling the data is as follows:

```
data hyper;
   infile 'hypertension.dat';
   input n1-n12;
   if _n_<4 then biofeed='P';
      else biofeed='A';
   if _n_ in(1,4) then drug='X';
   if _n_ in(2,5) then drug='Y';
   if _n_ in(3,6) then drug='Z';
   array nall {12} n1-n12;
   do i=1 to 12;
      if i>6 then diet='Y';
         else diet='N';
         bp=nall{i};
         cell=drug||biofeed||diet;
         output;
   end;
   drop i n1-n12;
run;
```

The 12 blood pressure readings per row, or line, of data are read into variables n1 - n12 and used to create 12 separate observations. The row and column positions in the data are used to determine the values of the factors in the design: drug, biofeed, and diet.

First, the input statement reads the 12 blood pressure values into variables n1 to n2. It uses list input, which assumes the data values to be separated by spaces.

The next group of statements uses the SAS automatic variable _n_ to determine which row of data is being processed and hence to set the values of drug and biofeed. Because six lines of data will be read, one line per iteration of the data step _n_ will increment from 1 to 6, corresponding to the line of data read with the input statement.

The key elements in splitting the one line of data into separate observations are the array, the do loop, and the output statement. The array statement defines an array by specifying the name of the array (nall here), the number of variables to be included in braces, and the list of variables to be included (n1 to n12 in this case).

In SAS, an array is a shorthand way of referring to a group of variables. In effect, it provides aliases for them so that each variable can be referred to using the name of the array and its position within the array in braces. For example, in this data step, n12 could be referred to as nall{12} or, when the variable i has the value 12 as nall{i}. However, the array only lasts for the duration of the data step in which it is defined.

The main purpose of an iterative do loop, like the one used here, is to repeat the statements between the do and the end a fixed number of times, with an index variable changing at each repetition. When used to process each of the variables in an array, the do loop should start with the index variable equal to 1 and end when it equals the number of variables in the array.

Within the do loop, in this example, the index variable i is first used to set the appropriate values for diet. Then a variable for the blood pressure reading (bp) is assigned one of the 12 values input. A character variable (cell) is formed by concatenating the values of the drug, biofeed, and diet variables. The double bar operator (||) concatenates character values.

The output statement writes an observation to the output data set with the current value of all variables. An output statement is not normally necessary because, without it an observation is automatically written out at the end of the data step. Putting an output statement within the do loop results in 12 observations being written to the data set.

Finally, the drop statement excludes the index variable i and n1 to n12 from the output data set because they are no longer needed.

As with any relatively complex data manipulation, it is wise to check that the results are as they should be, for example, by using proc print.

To begin the analysis, it is helpful to look at some summary statistics for each of the cells in the design.

```
proc tabulate data=hyper;
   class drug diet biofeed;
   var bp;
   table drug*diet*biofeed,
      bp*(mean std n);
run;
```

The **tabulate** procedure is useful for displaying descriptive statistics in a concise tabular form. The variables used in the table must first be declared in *either* a **class** statement or a **var** statement. Class variables are those used to divide the observations into groups. Those declared in the **var** statement (analysis variables) are those for which descriptive statistics are to be calculated. The first part of the **table** statement up to the comma specifies how the rows of the table are to be formed, and the remaining part specifies the columns. In this example, the rows comprise a hierarchical grouping of **biofeed** within **diet** within **drug**. The columns comprise the blood pressure mean and standard deviation and cell count for each of the groups. The resulting table is shown in Display 5.2. The differences between the standard deviations seen in this display may have implications for the analysis of variance of these data because one of the assumptions made is that observations in each cell come from populations with the same variance.

			bp		
			Mean	Std	N
drug	diet	biofeed			
X	N	A	188.00	10.86	6.00
		P	168.00	8.60	6.00
	Y	A	173.00	9.80	6.00
		P	169.00	14.82	6.00
Y	N	A	200.00	10.08	6.00
		P	204.00	12.68	6.00
	Y	A	187.00	14.01	6.00
		P	172.00	10.94	6.00
Z	N	A	209.00	14.35	6.00
		P	189.00	12.62	6.00
	Y	A	182.00	17.11	6.00
		P	173.00	11.66	6.00

Display 5.2

There are various ways in which the homogeneity of variance assumption can be tested. Here, the **hovtest** option of the **anova** procedure is used to apply Levene's test (Levene [1960]). The cell variable calculated above, which has 12 levels corresponding to the 12 cells of the design, is used:

```
proc anova data=hyper;
   class cell;
```

```
   model bp=cell;
   means cell / hovtest;
run;
```

The results are shown in Display 5.3. Concentrating on the results of Levene's test given in this display, we see that there is no formal evidence of heterogeneity of variance, despite the rather different observed standard deviations noted in Display 5.2.

```
                          The ANOVA Procedure

                        Class Level Information

 Class  Levels  Values

 cell       12  XAN XAY XPN XPY YAN YAY YPN YPY ZAN ZAY ZPN ZPY

                       Number of observations    72

                          The ANOVA Procedure

Dependent Variable: bp

                                    Sum of
     Source            DF          Squares   Mean Square   F Value   Pr > F

     Model             11      13194.00000    1199.45455      7.66   <.0001

     Error             60       9400.00000     156.66667

     Corrected Total   71      22594.00000


            R-Square    Coeff Var    Root MSE     bp Mean

            0.583960     6.784095    12.51666    184.5000


     Source    DF      Anova SS     Mean Square    F Value    Pr > F

       cell    11   13194.00000      1199.45455       7.66    <.0001
```

```
                    The ANOVA Procedure

        Levene's Test for Homogeneity of bp Variance
        ANOVA of Squared Deviations from Group Means

                    Sum of        Mean
Source      DF     Squares      Square    F Value    Pr > F

cell        11      180715     16428.6       1.01    0.4452
Error       60      971799     16196.6

                    The ANOVA Procedure

       Level of             -------------bp------------
       cell           N     Mean                Std Dev

       XAN            6     188.000000       10.8627805
       XAY            6     173.000000        9.7979590
       XPN            6     168.000000        8.6023253
       XPY            6     169.000000       14.8189068
       YAN            6     200.000000       10.0796825
       YAY            6     187.000000       14.0142784
       YPN            6     204.000000       12.6806940
       YPY            6     172.000000       10.9361785
       ZAN            6     209.000000       14.3527001
       ZAY            6     182.000000       17.1113997
       ZPN            6     189.000000       12.6174482
       ZPY            6     173.000000       11.6619038
```

Display 5.3

To apply the model specified in Eq. (5.1) to the hypertension data, proc anova can now be used as follows:

```
proc anova data=hyper;
  class diet drug biofeed;
  model bp=diet|drug|biofeed;
  means diet*drug*biofeed;
ods output means=outmeans;
run;
```

The anova procedure is specifically for balanced designs, that is, those with the same number of observations in each cell. (Unbalanced designs should be analysed using proc glm, as illustrated in a subsequent chapter.) The class statement specifies the classification variables, or factors. These may be numeric or character variables. The **model** statement specifies the dependent variable on the left-hand side of the equation and the effects (i.e., factors and their interactions) on the right-hand side of the equation. Main effects are specified by including the variable name and interactions by joining the variable names with an asterisk. Joining variable names with a bar is a shorthand way of specifying an interaction and all the lower-order interactions and main effects implied by it. Thus, the model statement above is equivalent to:

```
model bp=diet drug diet*drug biofeed diet*biofeed drug*biofeed
diet*drug*biofeed;
```

The order of the effects is determined by the expansion of the bar operator from left to right.

The **means** statement generates a table of cell means and the **ods output** statement specifies that this is to be saved in a SAS data set called **outmeans**.

The results are shown in Display 5.4. Here, it is the analysis of variance table that is of most interest. The **diet**, **biofeed**, and **drug** main effects are all significant beyond the 5% level. None of the first-order interactions are significant, but the three-way, second-order interaction of diet, drug, and biofeedback *is* significant. Just what does such an effect imply, and what are its implications for interpreting the analysis of variance results?

First, a significant second-order interaction implies that the first-order interaction between two of the variables differs in form or magnitude in the different levels of the remaining variable. Second, the presence of a significant second-order interaction means that there is little point in drawing conclusions about either the non-significant first-order interactions or the significant main effects. The effect of drug, for example, is not consistent for all combinations of diet and biofeedback. It would therefore be potentially misleading to conclude, on the basis of the significant main effect, anything about the specific effects of these three drugs on blood pressure.

The ANOVA Procedure

Class Level Information

Class	Levels	Values
diet	2	N Y
drug	3	X Y Z
biofeed	2	A P

```
                    Number of observations 72

                      The ANOVA Procedure

Dependent Variable: bp

                                 Sum of
 Source              DF         Squares    Mean Square   F Value    Pr > F

 Model               11     13194.00000     1199.45455      7.66    <.0001

 Error               60      9400.00000      156.66667

 Corrected Total     71     22594.00000

           R-Square    Coeff Var    Root MSE    bp Mean

           0.583960     6.784095    12.51666   184.5000

    Source                 DF     Anova SS  Mean Square  F Value  Pr > F

    diet                    1  5202.000000  5202.000000    33.20  <.0001
    drug                    2  3675.000000  1837.500000    11.73  <.0001
    diet*drug               2   903.000000   451.500000     2.88  0.0638
    biofeed                 1  2048.000000  2048.000000    13.07  0.0006
    diet*biofeed            1    32.000000    32.000000     0.20  0.6529
    drug*biofeed            2   259.000000   129.500000     0.83  0.4425
    diet*drug*biofeed       2  1075.000000   537.500000     3.43  0.0388

                      The ANOVA Procedure

     Level of    Level of    Level of            --------------bp-------------
     diet        drug        biofeed     N        Mean            Std Dev

     N           X           A           6    188.000000      10.8627805
     N           X           P           6    168.000000       8.6023253
     N           Y           A           6    200.000000      10.0796825
     N           Y           P           6    204.000000      12.6806940
     N           Z           A           6    209.000000      14.3527001
     N           Z           P           6    189.000000      12.6174482
     Y           X           A           6    173.000000       9.7979590
     Y           X           P           6    169.000000      14.8189068
     Y           Y           A           6    187.000000      14.0142784
     Y           Y           P           6    172.000000      10.9361785
     Y           Z           A           6    182.000000      17.1113997
     Y           Z           P           6    173.000000      11.6619038
```

Display 5.4

Understanding the meaning of the significant second-order interaction is facilitated by plotting some simple graphs. Here, the interaction plot of diet and biofeedback separately for each drug will help.

The cell means in the **outmeans** data set are used to produce interaction diagrams as follows:

```
proc print data=outmeans;
proc sort data=outmeans;
   by drug;

symbol1 i=join v=none l=2;
symbol2 i=join v=none l=1;

proc gplot data=outmeans;
   plot mean_bp*biofeed=diet ;
   by drug;
run;
```

First the **outmeans** data set is printed. The result is shown in Display 5.5. As well as checking the results, this also shows the name of the variable containing the means.

To produce separate plots for each drug, we use the **by** statement within **proc gplot**, but the data set must first be sorted by **drug**. **Plot** statements of the form **plot y*x=z** were introduced in Chapter 1 along with the **symbol** statement to change the plotting symbols used. We know that **diet** has two values, so we use two **symbol** statements to control the way in which the means for each value of **diet** are plotted. The **i** (**interpolation**) option specifies that the means are to be joined by lines. The **v** (**value**) option suppresses the plotting symbols because these are not needed and the **l** (**linetype**) option specifies different types of line for each diet. The resulting plots are shown in Displays 5.6 through 5.8. For drug X, the diet × biofeedback interaction plot indicates that diet has a negligible effect when biofeedback is given, but substantially reduces blood pressure when biofeedback is absent. For drug Y, the situation is essentially the reverse of that for drug X. For drug Z, the blood pressure difference when the diet is given and when it is not is approximately equal for both levels of biofeedback.

Obs	Effect	diet	drug	biofeed	N	Mean_bp	SD_bp
1	diet_drug_biofeed	N	X	A	6	188.000000	10.8627805
2	diet_drug_biofeed	N	X	P	6	168.000000	8.6023253
3	diet_drug_biofeed	N	Y	A	6	200.000000	10.0796825
4	diet_drug_biofeed	N	Y	P	6	204.000000	12.6806940
5	diet_drug_biofeed	N	Z	A	6	209.000000	14.3527001
6	diet_drug_biofeed	N	Z	P	6	189.000000	12.6174482
7	diet_drug_biofeed	Y	X	A	6	173.000000	9.7979590
8	diet_drug_biofeed	Y	X	P	6	169.000000	14.8189068
9	diet_drug_biofeed	Y	Y	A	6	187.000000	14.0142784
10	diet_drug_biofeed	Y	Y	P	6	172.000000	10.9361785
11	diet_drug_biofeed	Y	Z	A	6	182.000000	17.1113997
12	diet_drug_biofeed	Y	Z	P	6	173.000000	11.6619038

Display 5.5

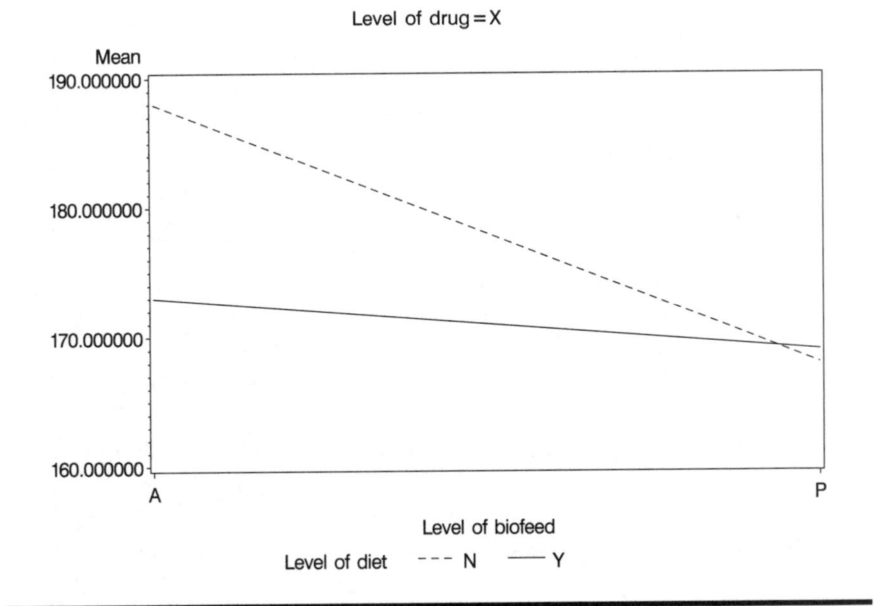

Display 5.6

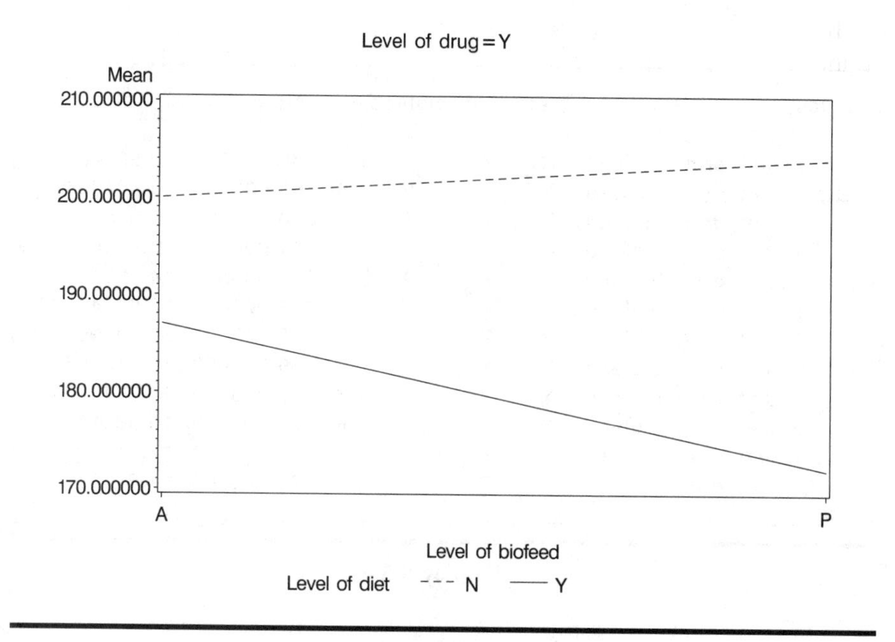

Display 5.7

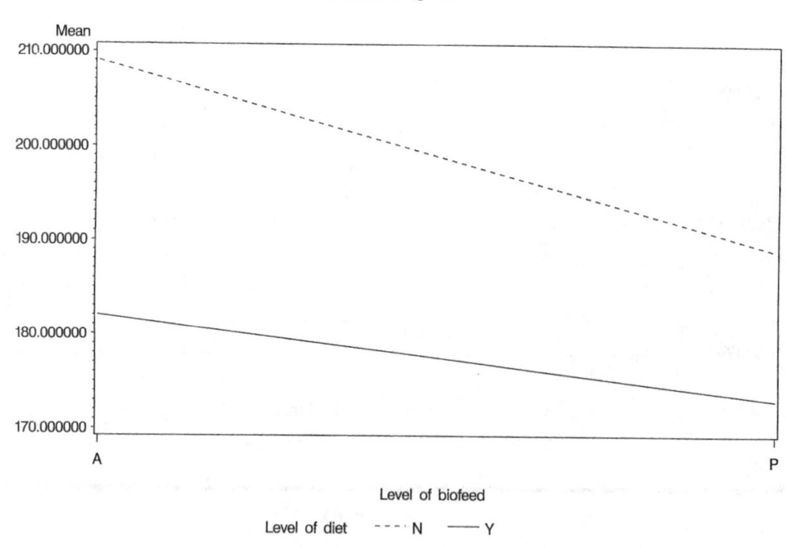

Display 5.8

In some cases, a significant high-order interaction may make it difficult to interpret the results from a factorial analysis of variance. In such cases, a transformation of the data may help. For example, we can analyze the log-transformed observations as follows:

```
data hyper;
   set hyper;
   logbp=log(bp);
run;

proc anova data=hyper;
   class diet drug biofeed;
   model logbp=diet|drug|biofeed;
run;
```

The data step computes the natural log of bp and stores it in a new variable logbp. The anova results for the transformed variable are given in Display 5.9.

```
                        The ANOVA Procedure

                      Class Level Information

                     Class        Levels    Values

                     diet              2    N Y

                     drug              3    X Y Z

                     biofeed           2    A P

                    Number of observations    72

                        The ANOVA Procedure

Dependent Variable: logbp

                                    Sum of
  Source              DF           Squares    Mean Square    F Value    Pr > F

  Model               11        0.37953489      0.03450317       7.46    <.0001

  Error               60        0.27754605      0.00462577

  Corrected Total     71        0.65708094
```

R-Square	Coeff Var	Root MSE	logbp Mean
0.577608	1.304662	0.068013	5.213075

Source	DF	Anova SS	Mean Square	F Value	Pr > F
diet	1	0.14956171	0.14956171	32.33	<.0001
drug	2	0.10706115	0.05353057	11.57	<.0001
diet*drug	2	0.02401168	0.01200584	2.60	0.0830
biofeed	1	0.06147547	0.06147547	13.29	0.0006
diet*biofeed	1	0.00065769	0.00065769	0.14	0.7075
drug*biofeed	2	0.00646790	0.00323395	0.70	0.5010
diet*drug*biofeed	2	0.03029929	0.01514965	3.28	0.0447

Display 5.9

Although the results are similar to those for the untransformed observations, the three-way interaction is now only marginally significant. If no substantive explanation of this interaction is forthcoming, it might be preferable to interpret the results in terms of the very significant main effects and fit a main-effects-only model to the log-transformed blood pressures. In addition, we can use Scheffe's multiple comparison test (Fisher and Van Belle, 1993) to assess which of the three drug means actually differ.

```
proc anova data=hyper;
   class diet drug biofeed;
   model logbp=diet drug biofeed;
   means drug / scheffe;
run;
```

The results are shown in Display 5.10. Each of the main effects is seen to be highly significant, and the grouping of means resulting from the application of Scheffe's test indicates that drug X produces lower blood pressures than the other two drugs, whose means do not differ.

The ANOVA Procedure

Class Level Information

Class	Levels	Values
diet	2	N Y
drug	3	X Y Z
biofeed	2	A P

Number of observations 72

The ANOVA Procedure

Dependent Variable: logbp

Source	DF	Sum of Squares	Mean Square	F Value	Pr > F
Model	4	0.31809833	0.07952458	15.72	<.0001
Error	67	0.33898261	0.00505944		
Corrected Total	71	0.65708094			

R-Square	Coeff Var	Root MSE	logbp Mean
0.484108	1.364449	0.071130	5.213075

Source	DF	Anova SS	Mean Square	F Value	Pr > F
diet	1	0.14956171	0.14956171	29.56	<.0001
drug	2	0.10706115	0.05353057	10.58	0.0001
biofeed	1	0.06147547	0.06147547	12.15	0.0009

The ANOVA Procedure

Scheffe's Test for logbp

NOTE: This test controls the Type I experimentwise error rate.

```
        Alpha                                    0.05
        Error Degrees of Freedom                   67
        Error Mean Square                    0.005059
        Critical Value of F                   3.13376
        Minimum Significant Difference         0.0514

   Means with the same letter are not significantly different.

         Scheffe Grouping        Mean     N    drug

                       A      5.24709    24    Y
                       A
                       A      5.23298    24    Z

                       B      5.15915    24    X
```

Display 5.10

Exercises

5.1 Compare the results given by Bonferonni *t*-tests and Duncan's multiple range test for the three drug means, with those given by Scheffe's test as reported in Display 5.10.

5.2 Produce box plots of the log-transformed blood pressures for (a) diet present, diet absent; (b) biofeedback present, biofeedback absent; and (c) drugs X, Y, and Z.

Chapter 6

Analysis of Variance II: School Attendance Amongst Australian Children

6.1 Description of Data

The data used in this chapter arise from a sociological study of Australian Aboriginal and white children reported by Quine (1975); they are given in Display 6.1. In this study, children of both sexes from four age groups (final grade in primary schools and first, second, and third form in secondary school) and from two cultural groups were used. The children in each age group were classified as slow or average learners. The response variable of interest was the number of days absent from school during the school year. (Children who had suffered a serious illness during the year were excluded.)

Cell	*Origin*	*Sex*	*Grade*	*Type*	*Days Absent*
1	A	M	F0	SL	2,11,14
2	A	M	F0	AL	5,5,13,20,22
3	A	M	F1	SL	6,6,15
4	A	M	F1	AL	7,14
5	A	M	F2	SL	6,32,53,57
6	A	M	F2	AL	14,16,16,17,40,43,46
7	A	M	F3	SL	12,15
8	A	M	F3	AL	8,23,23,28,34,36,38
9	A	F	F0	SL	3
10	A	F	F0	AL	5,11,24,45
11	A	F	F1	SL	5,6,6,9,13,23,25,32,53,54
12	A	F	F1	AL	5,5,11,17,19
13	A	F	F2	SL	8,13,14,20,47,48,60,81
14	A	F	F2	AL	2
15	A	F	F3	SL	5,9,7
16	A	F	F3	AL	0,2,3,5,10,14,21,36,40
17	N	M	F0	SL	6,17,67
18	N	M	F0	AL	0,0,2,7,11,12
19	N	M	F1	SL	0,0,5,5,5,11,17
20	N	M	F1	AL	3,3
21	N	M	F2	SL	22,30,36
22	N	M	F2	AL	8,0,1,5,7,16,27
23	N	M	F3	SL	12,15
24	N	M	F3	AL	0,30,10,14,27,41,69
25	N	F	F0	SL	25
26	N	F	F0	AL	10,11,20,33
27	N	F	F1	SL	5,7,0,1,5,5,5,5,7,11,15
28	N	F	F1	AL	5,14,6,6,7,28
29	N	F	F2	SL	0,5,14,2,2,3,8,10,12
30	N	F	F2	AL	1
31	N	F	F3	SL	8
32	N	F	F3	AL	1,9,22,3,3,5,15,18,22,37

Note: A, Aboriginal; N, non-Aboriginal; F, female; M, male; F0, primary; F1, first form; F2, second form; F3, third form; SL, slow learner; AL, average learner.

Display 6.1

6.2 Analysis of Variance Model

The basic design of the study is a 4 × 2 × 2 × 2 factorial. The usual model for y_{ijklm}, the number of days absent for the ith child in the jth sex group, the kth age group, the lth cultural group, and the mth learning group, is

$$y_{ijklm} = \mu + \alpha_j + \beta_k + \gamma_p + \delta_m + (\alpha\beta)_{jk} + (\alpha\gamma)_{jp} + (\alpha\delta)_{jm} + (\beta\gamma)_{kl}$$

$$+ (\beta\delta)_{km} + (\gamma\delta)_{lm} + (\alpha\beta\gamma)_{jkl} + (\alpha\beta\delta)_{jkm} + (\alpha\gamma\delta)_{jlm} + (\beta\gamma\delta)_{klm}$$

$$+ (\alpha\beta\gamma\delta)_{jklm} + \epsilon_{ijklm} \qquad (6.1)$$

where the terms represent main effects, first-order interactions of pairs of factors, second-order interactions of sets of three factors, and a third-order interaction for all four factors. (The parameters must be constrained in some way to make the model identifiable. Most common is to require they sum to zero over any subscript.) The ϵ_{ijklm} represent random error terms assumed to be normally distributed with mean zero and variance σ^2.

The unbalanced nature of the data in Display 6.1 (there are different numbers of observations for the different combinations of factors) presents considerably more problems than encountered in the analysis of the balanced factorial data in the previous chapter. The main difficulty is that when the data are unbalanced, there is no unique way of finding a "sums of squares" corresponding to each main effect and each interaction because these effects are no longer independent of one another. It is now no longer possible to partition the total variation in the response variable into non-overlapping or orthogonal sums of squares representing factor main effects and factor interactions. For example, there is a proportion of the variance of the response variable that can be attributed to (explained by) either sex or age group and, consequently, sex and age group together explain less of the variation of the response than the sum of which each explains alone. The result of this is that the sums of squares that can be attributed to a factor depends on which factors have already been allocated a sums of squares; that is, the sums of squares of factors and their interactions depend on the order in which they are considered.

The dependence between the factor variables in an unbalanced factorial design and the consequent lack of uniqueness in partitioning the variation in the response variable has led to a great deal of confusion regarding what is the most appropriate way to analyse such designs. The issues are not straightforward and even statisticians (yes, even statisticians!) do not wholly agree on the most suitable method of analysis for all situations, as is witnessed by the discussion following the papers of Nelder (1977) and Aitkin (1978).

Essentially the discussion over the analysis of unbalanced factorial designs has involved the question of what type of sums of squares should be used. Basically there are three possibilities; but only two are considered here, and these are illustrated for a design with two factors.

6.2.1 Type I Sums of Squares

These sums of squares represent the effect of adding a term to an existing model in one particular order. Thus, for example, a set of Type I sums of squares such as:

Source	*Type I SS*
A	SSA
B	SSB|A
AB	SSAB|A,B

essentially represent a comparison of the following models:

SSAB|A,B	Model including an interaction and main effects with one including only main effects
SSB|A	Model including both main effects, but no interaction, with one including only the main effect of factor A
SSA	Model containing only the A main effect with one containing only the overall mean

The use of these sums of squares in a series of tables in which the effects are considered in different orders (see later) will often provide the most satisfactory way of answering the question as to which model is most appropriate for the observations.

6.2.2 Type III Sums of Squares

Type III sums of squares represent the contribution of each term to a model including all other possible terms. Thus, for a two-factor design, the sums of squares represent the following:

Source	*Type III SS*
A	SSA|B,AB
B	SSB|A,AB
AB	SSAB|A,B

(SAS also has a Type IV sum of squares, which is the same as Type III unless the design contains empty cells.)

In a balanced design, Type I and Type III sums of squares are equal; but for an unbalanced design, they are not and there have been numerous discussions regarding which type is most appropriate for the analysis of such designs. Authors such as Maxwell and Delaney (1990) and Howell (1992) strongly recommend the use of Type III sums of squares and these are the default in SAS. Nelder (1977) and Aitkin (1978), however, are strongly critical of "correcting" main effects sums of squares for an interaction term involving the corresponding main effect; their criticisms are based on both theoretical and pragmatic grounds. The arguments are relatively subtle but in essence go something like this:

- When fitting models to data, the principle of *parsimony* is of critical importance. In choosing among possible models, we do not adopt complex models for which there is no empirical evidence.
- Thus, if there is no convincing evidence of an AB interaction, we do not retain the term in the model. Thus, additivity of A and B is assumed unless there is convincing evidence to the contrary.
- So the argument proceeds that Type III sum of squares for A in which it is adjusted for AB makes no sense.
- First, if the interaction term is necessary in the model, then the experimenter will usually want to consider simple effects of A at each level of B separately. A test of the hypothesis of no A main effect would not usually be carried out if the AB interaction is significant.
- If the AB interaction is not significant, then adjusting for it is of no interest, and causes a substantial loss of power in testing the A and B main effects.

(The issue does not arise so clearly in the balanced case, for there the sum of squares for A say is independent of whether or not interaction is assumed. Thus, in deciding on possible models for the data, the interaction term is not included unless it has been shown to be necessary, in which case tests on main effects involved in the interaction are not carried out; or if carried out, not interpreted — see biofeedback example in Chapter 5.)

The arguments of Nelder and Aitkin against the use of Type III sums of squares are powerful and persuasive. Their recommendation to use Type I sums of squares, considering effects in a number of orders, as the most suitable way in which to identify a suitable model for a data set is also convincing and strongly endorsed by the authors of this book.

6.3 Analysis Using SAS

It is assumed that the data are in an ASCII file called **ozkids.dat** in the current directory and that the values of the factors comprising the design are separated by tabs, whereas those recoding days of absence for the subjects within each cell are separated by commas, as in Display 6.1. The data can then be read in as follows:

```
data ozkids;
   infile 'ozkids.dat' dlm=' ,' expandtabs missover;
   input cell origin $ sex $ grade $ type $ days @;
      do until (days=.);
      output;
      input days @;
   end;
   input;
run;
```

The **expandtabs** option on the **infile** statement converts tabs to spaces so that list input can be used to read the tab-separated values. To read the comma-separated values in the same way, the **delimiter** option (abbreviated **dlm**) specifies that both spaces and commas are delimiters. This is done by including a space and a comma in quotes after **dlm=**. The **missover** option prevents SAS from reading the next line of data in the event that an **input** statement requests more data values than are contained in the current line. Missing values are assigned to the variable(s) for which there are no corresponding data values. To illustrate this with an example, suppose we have an **input** statement **input x1-x7;**. If a line of data only contains five numbers, by default SAS will go to the next line of data to read data values for **x6** and **x7**. This is not usually what is intended; so when it happens, there is a warning message in the log: "SAS went to a new line when INPUT statement reached past the end of a line." With the **missover** option, SAS would not go to a new line but **x6** and **x7** would have missing values. Here we utilise this to determine when all the values for days of absence from school have been read.

The **input** statement reads the cell number, the factors in the design, and the days absent for the first observation in the cell. The trailing **@** at the end of the statement holds the data line so that more data can be read from it by subsequent **input** statements. The statements between the **do until** and the following **end** are repeatedly executed until the **days** variable has a missing value. The **output** statement creates an observation in the output data set. Then another value of **days** is read, again holding the data line with a trailing **@**. When all the values from the line have been read, and output as observations, the days variable is assigned a missing value and the do until loop finishes. The following input statement then releases the data line so that the next line of data from the input file can be read.

For unbalanced designs, the glm procedure should be used rather than proc anova. We begin by fitting main-effects-only models for different orders of main effects.

```
proc glm data=ozkids;
   class origin sex grade type;
   model days=origin sex grade type /ss1 ss3;

proc glm data=ozkids;
   class origin sex grade type;
   model days=grade sex type origin /ss1;

proc glm data=ozkids;
   class origin sex grade type;
   model days=type sex origin grade /ss1;

proc glm data=ozkids;
   class origin sex grade type;
   model days=sex origin type grade /ss1;
run;
```

The class statement specifies the classification variables, or factors. These can be numeric or character variables. The model statement specifies the dependent variable on the left-hand side of the equation and the effects (i.e., factors and their interactions) on the right-hand side of the equation. Main effects are specified by including the variable name.

The options in the model statement in the first glm step specify that both Type I and Type III sums of squares are to be output. The subsequent proc steps repeat the analysis, varying the order of the effects; but because Type III sums of squares are invariant to the order, only Type I sums of squares are requested. The output is shown in Display 6.2. Note that when a main effect is ordered last, the corresponding Type I sum of squares is the same as the Type III sum of squares for the factor. In fact, when dealing with a main-effects only model, the Type III sums of squares *can* legitimately be used to identify the most important effects. Here, it appears that origin and grade have the most impact on the number of days a child is absent from school.

The GLM Procedure

Class Level Information

Class	Levels	Values
origin	2	A N
sex	2	F M
grade	4	F0 F1 F2 F3
type	2	AL SL

Number of observations 154

The GLM Procedure

Dependent Variable: days

Source	DF	Sum of Squares	Mean Square	F Value	Pr > F
Model	6	4953.56458	825.59410	3.60	0.0023
Error	147	33752.57179	229.60933		
Corrected Total	153	38706.13636			

R-Square	Coeff Var	Root MSE	days Mean
0.127979	93.90508	15.15287	16.13636

Source	DF	Type I SS	Mean Square	F Value	Pr > F
origin	1	2645.652580	2645.652580	11.52	0.0009
sex	1	338.877090	338.877090	1.48	0.2264
grade	3	1837.020006	612.340002	2.67	0.0500
type	1	132.014900	132.014900	0.57	0.4495

Source	DF	Type III SS	Mean Square	F Value	Pr > F
origin	1	2403.606653	2403.606653	10.47	0.0015
sex	1	185.647389	185.647389	0.81	0.3700
grade	3	1917.449682	639.149894	2.78	0.0430
type	1	132.014900	132.014900	0.57	0.4495

```
                         The GLM Procedure

                      Class Level Information

                   Class     Levels    Values

                   origin       2      A N

                   sex          2      F M

                   grade        4      F0 F1 F2 F3

                   type         2      AL SL

                   Number of observations    154

                         The GLM Procedure

Dependent Variable: days

                                   Sum of
   Source               DF        Squares  Mean Square  F Value  Pr > F

   Model                 6     4953.56458    825.59410     3.60  0.0023

   Error               147    33752.57179    229.60933

   Corrected Total     153    38706.13636


           R-Square    Coeff Var    Root MSE    days Mean

           0.127979     93.90508    15.15287     16.13636


    Source    DF      Type I SS    Mean Square    F Value    Pr > F

    grade      3    2277.172541     759.057514       3.31    0.0220
    sex        1     124.896018     124.896018       0.54    0.4620
    type       1     147.889364     147.889364       0.64    0.4235
    origin     1    2403.606653    2403.606653      10.47    0.0015
```

The GLM Procedure

Class Level Information

Class	Levels	Values
origin	2	A N
sex	2	F M
grade	4	F0 F1 F2 F3
type	2	AL SL

Number of observations 154

The GLM Procedure

Dependent Variable: days

Source	DF	Sum of Squares	Mean Square	F Value	Pr > F
Model	6	4953.56458	825.59410	3.60	0.0023
Error	147	33752.57179	229.60933		
Corrected Total	153	38706.13636			

R-Square	Coeff Var	Root MSE	days Mean
0.127979	93.90508	15.15287	16.13636

Source	DF	Type I SS	Mean Square	F Value	Pr > F
type	1	19.502391	19.502391	0.08	0.7711
sex	1	336.215409	336.215409	1.46	0.2282
origin	1	2680.397094	2680.397094	11.67	0.0008
grade	3	1917.449682	639.149894	2.78	0.0430

```
                         The GLM Procedure

                      Class Level Information

                    Class      Levels    Values

                    origin        2      A N

                    sex           2      F M

                    grade         4      F0 F1 F2 F3

                    type          2      AL SL

                    Number of observations    154

                         The GLM Procedure

Dependent Variable: days

                                     Sum of
  Source               DF          Squares   Mean Square  F Value   Pr > F

  Model                 6       4953.56458     825.59410     3.60   0.0023

  Error               147      33752.57179     229.60933

  Corrected Total     153      38706.13636


           R-Square     Coeff Var     Root MSE     days Mean

           0.127979      93.90508     15.15287      16.13636


     Source     DF        Type I SS    Mean Square    F Value    Pr > F

     sex         1       308.062554     308.062554       1.34    0.2486
     origin      1      2676.467116    2676.467116      11.66    0.0008
     type        1        51.585224      51.585224       0.22    0.6362
     grade       3      1917.449682     639.149894       2.78    0.0430
```

Display 6.2

Next we fit a full factorial model to the data as follows:

```
proc glm data=ozkids;
   class origin sex grade type;
   model days=origin sex grade type origin|sex|grade|type /ss1
ss3;
run;
```

Joining variable names with a bar is a shorthand way of specifying an interaction and all the lower-order interactions and main effects implied by it. This is useful not only to save typing but to ensure that relevant terms in the model are not inadvertently omitted. Here we have explicitly specified the main effects so that they are entered before any interaction terms when calculating Type I sums of squares.

The output is shown in Display 6.3. Note first that the only Type I and Type III sums of squares that agree are those for the **origin * sex * grade * type** interaction. Now consider the origin main effect. The Type I sum of squares for origin is "corrected" only for the mean because it appears first in the **proc glm** statement. The effect is highly significant. But using Type III sums of squares, in which the origin effect is corrected for all other main effects and interactions, the corresponding F value has an associated P-value of 0.2736. Now origin is judged nonsignificant, but this may simply reflect the loss of power after "adjusting" for a lot of relatively unimportant interaction terms.

Arriving at a final model for these data is not straightforward (see Aitkin [1978] for some suggestions), and the issue is not pursued here because the data set will be the subject of further analyses in Chapter 9. However, some of the exercises encourage readers to try some alternative analyses of variance.

```
                    The GLM Procedure

                 Class Level Information

            Class    Levels    Values

            origin        2    A N

            se            2    F M

            grade         4    F0 F1 F2 F3

            type          2    AL SL

            Number of observations    154
```

```
                          The GLM Procedure

Dependent Variable: days

                                  Sum of
 Source              DF          Squares   Mean Square   F Value   Pr > F

 Model               31      15179.41930     489.65869      2.54   0.0002

 Error              122      23526.71706     192.84194

 Corrected Total    153      38706.13636


            R-Square   Coeff Var   Root MSE   days Mean

            0.392171   86.05876    13.88675    16.13636


  Source                  DF     Type I SS   Mean Square  F Value  Pr > F

  origin                   1   2645.652580   2645.652580    13.72  0.0003
  sex                      1    338.877090    338.877090     1.76  0.1874
  grade                    3   1837.020006    612.340002     3.18  0.0266
  type                     1    132.014900    132.014900     0.68  0.4096
  origin*sex               1    142.454554    142.454554     0.74  0.3918
  origin*grade             3   3154.799178   1051.599726     5.45  0.0015
  sex*grade                3   2009.479644    669.826548     3.47  0.0182
  origin*sex*grade         3    226.309848     75.436616     0.39  0.7596
  origin*type              1     38.572890     38.572890     0.20  0.6555
  sex*type                 1     69.671759     69.671759     0.36  0.5489
  origin*sex*type          1    601.464327    601.464327     3.12  0.0799
  grade*type               3   2367.497717    789.165906     4.09  0.0083
  origin*grade*type        3    887.938926    295.979642     1.53  0.2089
  sex*grade*type           3    375.828965    125.276322     0.65  0.5847
  origi*sex*grade*type     3    351.836918    117.278973     0.61  0.6109
```

Source	DF	Type III SS	Mean Square	F Value	Pr > F
origin	1	233.201138	233.201138	1.21	0.2736
sex	1	344.037143	344.037143	1.78	0.1841
grade	3	1036.595762	345.531921	1.79	0.1523
type	1	181.049753	181.049753	0.94	0.3345
origin*sex	1	3.261543	3.261543	0.02	0.8967
origin*grade	3	1366.765758	455.588586	2.36	0.0746
sex*grade	3	1629.158563	543.052854	2.82	0.0420
origin*sex*grade	3	32.650971	10.883657	0.06	0.9823
origin*type	1	55.378055	55.378055	0.29	0.5930
sex*type	1	1.158990	1.158990	0.01	0.9383
origin*sex*type	1	337.789437	337.789437	1.75	0.1881
grade*type	3	2037.872725	679.290908	3.52	0.0171
origin*grade*type	3	973.305369	324.435123	1.68	0.1743
sex*grade*type	3	410.577832	136.859277	0.71	0.5480
origi*sex*grade*type	3	351.836918	117.278973	0.61	0.6109

Display 6.3

Exercises

6.1 Investigate simpler models for the data used in this chapter by dropping interactions or sets of interactions from the full factorial model fitted in the text. Try several different orders of effects.

6.2 The outcome for the data in this chapter — number of days absent — is a count variable. Consequently, assuming normally distributed errors may not be entirely appropriate, as we will see in Chapter 9. Here, however, we might deal with this potential problem by way of a transformation. One possibility is a log transformation. Investigate this possibility.

6.3 Find a table of cell means and standard deviations for the data used in this chapter.

6.4 Construct a normal probability plot of the residuals from fitting a main-effects-only model to the data used in this chapter. Comment on the results.

Chapter 7

Analysis of Variance of Repeated Measures: Visual Acuity

7.1 Description of Data

The data used in this chapter are taken from Table 397 of *SDS*. They are reproduced in Display 7.1. Seven subjects had their response times measured when a light was flashed into each eye through lenses of powers 6/6, 6/18, 6/36, and 6/60. Measurements are in milliseconds, and the question of interest was whether or not the response time varied with lens strength. (A lens of power a/b means that the eye will perceive as being at "a" feet an object that is actually positioned at "b" feet.)

7.2 Repeated Measures Data

The observations in Display 7.1 involve *repeated measures*. Such data arise often, particularly in the behavioural sciences and related disciplines, and involve recording the value of a response variable for each subject under more than one condition and/or on more than one occasion.

	Visual Acuity and Lens Strength							
	Left Eye				Right Eye			
Subject	*6/6*	*6/18*	*6/36*	*6/60*	*6/6*	*6/18*	*6/36*	*6/60*
1	116	119	116	124	120	117	114	122
2	110	110	114	115	106	112	110	110
3	117	118	120	120	120	120	120	124
4	112	116	115	113	115	116	116	119
5	113	114	114	118	114	117	116	112
6	119	115	94	116	100	99	94	97
7	110	110	105	118	105	105	115	115

Display 7.1

Researchers typically adopt the repeated measures paradigm as a means of reducing error variability and/or as the natural way of measuring certain phenomena (e.g., developmental changes over time, learning and memory tasks, etc). In this type of design, the effects of experimental factors giving rise to the repeated measures are assessed relative to the average response made by a subject on all conditions or occasions. In essence, each subject serves as his or her own control and, accordingly, variability due to differences in average responsiveness of the subjects is eliminated from the extraneous error variance. A consequence of this is that the power to detect the effects of within-subjects experimental factors is increased compared to testing in a between-subjects design.

Unfortunately, the advantages of a repeated measures design come at a cost, and that cost is the probable lack of independence of the repeated measurements. Observations made under different conditions involving the same subjects will very likely be correlated rather than independent. This violates one of the assumptions of the analysis of variance procedures described in Chapters 5 and 6, and accounting for the dependence between observations in a repeated measures designs requires some thought. (In the visual acuity example, only within-subject factors occur; and it is possible — indeed likely — that the lens strengths under which a subject was observed were given in random order. However, in examples where *time* is the single within-subject factor, randomisation is not, of course, an option. This makes the type of study in which subjects are simply observed over time rather different from other repeated measures designs, and they are often given a different label — *longitudinal designs*. Owing to their different nature, we consider them specifically later in Chapters 10 and 11.)

7.3 Analysis of Variance for Repeated Measures Designs

Despite the lack of independence of the observations made within subjects in a repeated measures design, it remains possible to use relatively straightforward analysis of variance procedures to analyse the data if three particular assumptions about the observations are valid; that is

1. *Normality*: the data arise from populations with normal distributions.
2. *Homogeneity of variance*: the variances of the assumed normal distributions are equal.
3. *Sphericity*: the variances of the differences between all pairs of the repeated measurements are equal. This condition implies that the correlations between pairs of repeated measures are also equal, the so-called *compound symmetry pattern*.

It is the third assumption that is most critical for the validity of the analysis of variance *F*-tests. When the sphericity assumption is not regarded as likely, there are two alternatives to a simple analysis of variance: the use of *correction factors* and *multivariate analysis of variance*. All three possibilities will be considered in this chapter.

We begin by considering a simple model for the visual acuity observations, y_{ijk}, where y_{ijk} represents the reaction time of the *i*th subject for eye *j* and lens strength *k*. The model assumed is

$$y_{ijk} = \mu + \alpha_j + \beta_k + (\alpha\beta)_{jk} + \gamma_i + (\gamma\alpha)_{ij} + (\gamma\beta)_{ik} + (\gamma\alpha\beta)_{ijk} + \epsilon_{ijk} \quad (7.1)$$

where α_j represents the effect of eye *j*, β_k is the effect of the *k*th lens strength, and $(\alpha\beta)_{jk}$ is the eye × lens strength interaction. The term γ_i is a constant associated with subject *i* and $(\gamma\alpha)_{ij}$, $(\gamma\beta)_{ik}$, and $(\gamma\alpha\beta)_{ijk}$ represent interaction effects of subject *i* with each factor and their interaction. The terms α_j, β_k, and $(\alpha\beta)_{jk}$ are assumed to be fixed effects, but the subject and error terms are assumed to be random variables from normal distributions with zero means and variances specific to each term. This is an example of a *mixed model*.

Equal correlations between the repeated measures arise as a consequence of the subject effects in this model; and if this structure is valid, a relatively straightforward analysis of variance of the data can be used. However, when the investigator thinks the assumption of equal correlations is too strong, there are two alternatives that can be used:

1. *Correction factors.* Box (1954) and Greenhouse and Geisser (1959) considered the effects of departures from the sphericity assumption in a repeated measures analysis of variance. They demonstrated that the extent to which a set of repeated measures departs from the sphericity assumption can be summarised in terms of a parameter $\in$, which is a function of the variances and covariances of the repeated measures. And an estimate of this parameter can be used to decrease the degrees of freedom of *F*-tests for the within-subjects effect to account for deviation from sphericity. In this way, larger *F*-values will be needed to claim statistical significance than when the correction is not used, and thus the increased risk of falsely rejecting the null hypothesis is removed. Formulae for the correction factors are given in Everitt (2001).
2. *Multivariate analysis of variance.* An alternative to the use of correction factors in the analysis of repeated measures data when the sphericity assumption is judged to be inappropriate is to use multivariate analysis of variance. The advantage is that no assumptions are now made about the pattern of correlations between the repeated measurements. A disadvantage of using MANOVA for repeated measures is often stated to be the technique's relatively low power when the assumption of compound symmetry is actually valid. However, Davidson (1972) shows that this is really only a problem with small sample sizes.

7.4 Analysis Using SAS

Assuming the ASCII file 'visual.dat' is in the current directory, the data can be read in as follows:

```
data vision;
   infile 'visual.dat' expandtabs;
   input idno x1-x8;
run;
```

The data are tab separated and the expandtabs option on the infile statement converts the tabs to spaces as the data are read, allowing a simple list input statement to be used.

The glm procedure is used for the analysis:

```
proc glm data=vision;
   model x1-x8= / nouni;
   repeated eye 2, strength 4 /summary;
run;
```

The eight repeated measures per subject are all specified as response variables in the **model** statement and thus appear on the left-hand side of the equation. There are no between-subjects factors in the design, so the right-hand side of the equation is left blank. Separate univariate analyses of the eight measures are of no interest and thus the **nouni** option is included to suppress them.

The **repeated** statement specifies the within-subjects factor structure. Each factor is given a name, followed by the number of levels it has. Factor specifications are separated by commas. The order in which they occur implies a data structure in which the factors are nested from right to left; in this case, one where lens strength is nested within eye. It is also possible to specify the type of contrasts to be used for each within-subjects factor. The default is to contrast each level of the factor with the previous. The **summary** option requests ANOVA tables for each contrast.

The output is shown in Display 7.2. Concentrating first on the univariate tests, we see that none of the effects — eye, strength, or eye × strength — are significant, and this is so whichever P-value is used, unadjusted, Greenhouse and Geisser (G-G) adjusted, or Huynh-Feldt (H-F) adjusted. However, the multivariate tests have a different story to tell; now the strength factor is seen to be highly significant.

Because the strength factor is on an ordered scale, we might investigate it further using orthogonal polynomial contrasts, here a linear, quadratic, and cubic contrast.

```
                          The GLM Procedure

                       Number of observations 7

                          The GLM Procedure
                Repeated Measures Analysis of Variance

                 Repeated Measures Level Information

  Dependent Variable    x1   x2   x3   x4   x5   x6   x7   x8

  Level of eye           1    1    1    1    2    2    2    2
  Level of strength      1    2    3    4    1    2    3    4

 Manova Test Criteria and Exact F Statistics for the Hypothesis of no
                              eye Effect
                  H = Type III SSCP Matrix for eye
                       E = Error SSCP Matrix

                  S=1        M=-0.5        N=2
```

Statistic	Value	F Value	Num DF	Den DF	Pr > F
Wilks' Lambda	0.88499801	0.78	1	6	0.4112
Pillai's Trace	0.11500199	0.78	1	6	0.4112
Hotelling-Lawley Trace	0.12994604	0.78	1	6	0.4112
Roy's Greatest Root	0.12994604	0.78	1	6	0.4112

Manova Test Criteria and Exact F Statistics for the Hypothesis of no strength Effect
H = Type III SSCP Matrix for strength
E = Error SSCP Matrix

S=1 M=0.5 N=1

Statistic	Value	F Value	Num DF	Den DF	Pr > F
Wilks' Lambda	0.05841945	21.49	3	4	0.0063
Pillai's Trace	0.94158055	21.49	3	4	0.0063
Hotelling-Lawley Trace	16.11758703	21.49	3	4	0.0063
Roy's Greatest Root	16.11758703	21.49	3	4	0.0063

Manova Test Criteria and Exact F Statistics for the Hypothesis of no eye*strength Effect
H = Type III SSCP Matrix for eye*strength
E = Error SSCP Matrix

S=1 M=0.5 N=1

Statistic	Value	F Value	Num DF	Den DF	Pr > F
Wilks' Lambda	0.70709691	0.55	3	4	0.6733
Pillai's Trace	0.29290309	0.55	3	4	0.6733
Hotelling-Lawley Trace	0.41423331	0.55	3	4	0.6733
Roy's Greatest Root	0.41423331	0.55	3	4	0.6733

The GLM Procedure
Repeated Measures Analysis of Variance
Univariate Tests of Hypotheses for Within Subject Effects

Source	DF	Type III SS	Mean Square	F Value	Pr > F
eye	1	46.4464286	46.4464286	0.78	0.4112
Error(eye)	6	357.4285714	59.5714286		

Source	DF	Type III SS	Mean Square	F Value	Pr > F	Adj Pr > F G – G	Adj Pr > F H - F
strength	3	140.7678571	46.9226190	2.25	0.1177	0.1665	0.1528
Error(strength)	18	375.8571429	20.8809524				

Greenhouse-Geisser Epsilon 0.4966
Huynh-Feldt Epsilon 0.6229

Source	DF	Type III SS	Mean Square	F Value	Pr > F	Adj Pr > F G – G	Adj Pr > F H - F
eye*strength	3	40.6250000	13.5416667	1.06	0.3925	0.3700	0.3819
Error(eye*strength)	18	231.0000000	12.8333333				

Greenhouse-Geisser Epsilon 0.5493
Huynh-Feldt Epsilon 0.7303

The GLM Procedure
Repeated Measures Analysis of Variance
Analysis of Variance of Contrast Variables

eye_N represents the contrast between the nth level of eye and the last

Contrast Variable: eye_1

Source	DF	Type III SS	Mean Square	F Value	Pr > F
Mean	1	371.571429	371.571429	0.78	0.4112
Error	6	2859.428571	476.571429		

The GLM Procedure
Repeated Measures Analysis of Variance
Analysis of Variance of Contrast Variables

strength_N represents the contrast between the nth level of strength and the last

Contrast Variable: strength_1

Source	DF	Type III SS	Mean Square	F Value	Pr > F
Mean	1	302.2857143	302.2857143	5.64	0.0552
Error	6	321.7142857	53.6190476		

```
Contrast Variable: strength_2

      Source    DF     Type III SS    Mean Square    F Value    Pr > F

      Mean       1     175.0000000    175.0000000       3.55    0.1086
      Error      6     296.0000000     49.3333333

Contrast Variable: strength_3

      Source    DF     Type III SS    Mean Square    F Value    Pr > F

      Mean       1     514.2857143    514.2857143       5.57    0.0562
      Error      6     553.7142857     92.2857143

                        The GLM Procedure
              Repeated Measures Analysis of Variance
             Analysis of Variance of Contrast Variables

eye_N represents the contrast between the nth level of eye and the last
strength_N represents the contrast between the nth level of strength and the
last

Contrast Variable: eye_1*strength_1

      Source    DF     Type III SS    Mean Square    F Value    Pr > F

      Mean       1      9.14285714     9.14285714       0.60    0.4667
      Error      6     90.85714286    15.14285714

Contrast Variable: eye_1*strength_2

      Source    DF     Type III SS    Mean Square    F Value    Pr > F

      Mean       1      11.5714286     11.5714286       0.40    0.5480
      Error      6     171.4285714     28.5714286

Contrast Variable: eye_1*strength_3

      Source    DF     Type III SS    Mean Square    F Value    Pr > F

      Mean       1     146.2857143    146.2857143       1.79    0.2291
      Error      6     489.7142857     81.6190476
```

Display 7.2

Polynomial contrasts for lens strength can be obtained by re-submitting the previous **glm** step with the following repeated statement:

```
repeated eye 2, strength 4 (1 3 6 10) polynomial /summary;
```

The specification of the lens strength factor has been expanded: numeric values for the four levels of lens strength have been specified in parentheses and orthogonal polynomial contrasts requested. The values specified will be used as spacings in the calculation of the polynomials.

The edited results are shown in Display 7.3. None of the contrasts are significant, although it must be remembered that the sample size here is small, so that the tests are not very powerful. The difference between the multivariate and univariate tests might also be due to the covariance structure departing from the univariate assumption of compound symmetry. Interested readers might want to examine this possibility.

```
                          The GLM Procedure

                     Number of observations 7

                          The GLM Procedure
                Repeated Measures Analysis of Variance

                 Repeated Measures Level Information

 Dependent Variable   x1   x2   x3   x4   x5   x6   x7   x8

       Level of eye    1    1    1    1    2    2    2    2
  Level of strength    1    3    6   10    1    3    6   10

 Manova Test Criteria and Exact F Statistics for the Hypothesis of no
                              eye Effect
                   H = Type III SSCP Matrix for eye
                        E = Error SSCP Matrix

                     S=1        M=-0.5        N=2

 Statistic                      Value  F Value  Num DF  Den DF  Pr > F

 Wilks' Lambda             0.88499801     0.78       1       6  0.4112
 Pillai's Trace            0.11500199     0.78       1       6  0.4112
 Hotelling-Lawley Trace    0.12994604     0.78       1       6  0.4112
 Roy's Greatest Root       0.12994604     0.78       1       6  0.4112

 Manova Test Criteria and Exact F Statistics for the Hypothesis of no
                           strength Effect
                H = Type III SSCP Matrix for strength
                        E = Error SSCP Matrix

                     S=1        M=0.5        N=1
```

Statistic	Value	F Value	Num DF	Den DF	Pr > F
Wilks' Lambda	0.05841945	21.49	3	4	0.0063
Pillai's Trace	0.94158055	21.49	3	4	0.0063
Hotelling-Lawley Trace	16.11758703	21.49	3	4	0.0063
Roy's Greatest Root	16.11758703	21.49	3	4	0.0063

Manova Test Criteria and Exact F Statistics for the Hypothesis of no eye*
strength Effect
H = Type III SSCP Matrix for eye*strength
E = Error SSCP Matrix

S=1 M=0.5 N=1

Statistic	Value	F Value	Num DF	Den DF	Pr > F
Wilks' Lambda	0.70709691	0.55	3	4	0.6733
Pillai's Trace	0.29290309	0.55	3	4	0.6733
Hotelling-Lawley Trace	0.41423331	0.55	3	4	0.6733
Roy's Greatest Root	0.41423331	0.55	3	4	0.6733

The GLM Procedure
Repeated Measures Analysis of Variance
Univariate Tests of Hypotheses for Within Subject Effects

Source	DF	Type III SS	Mean Square	F Value	Pr > F
eye	1	46.4464286	46.4464286	0.78	0.4112
Error(eye)	6	357.4285714	59.5714286		

						Adj Pr > F	
Source	DF	Type III SS	Mean Square	F Value	Pr > F	G - G	H - F
strength	3	140.7678571	46.9226190	2.25	0.1177	0.1665	0.1528
Error(strength)	18	375.8571429	20.8809524				

Greenhouse-Geisser Epsilon 0.4966
Huynh-Feldt Epsilon 0.6229

						Adj Pr > F	
Source	DF	Type III SS	Mean Square	F Value	Pr > F	G - G	H - F
eye*strength	3	40.6250000	13.5416667	1.06	0.3925	0.3700	0.3819
Error(eye*strength)	18	231.0000000	12.8333333				

```
            Greenhouse-Geisser Epsilon    0.5493
            Huynh-Feldt Epsilon           0.7303

                       The GLM Procedure
             Repeated Measures Analysis of Variance
            Analysis of Variance of Contrast Variables

eye_N represents the contrast between the nth level of eye and the last

Contrast Variable: eye_1

    Source    DF    Type III SS    Mean Square    F Value    Pr > F

    Mean       1     371.571429     371.571429       0.78    0.4112
    Error      6    2859.428571     476.571429

                       The GLM Procedure
             Repeated Measures Analysis of Variance
            Analysis of Variance of Contrast Variables

strength_N represents the nth degree polynomial contrast for strength

Contrast Variable: strength_1

    Source    DF    Type III SS    Mean Square    F Value    Pr > F

    Mean       1    116.8819876    116.8819876       2.78    0.1468
    Error      6    252.6832298     42.1138716

Contrast Variable: strength_2

    Source    DF    Type III SS    Mean Square    F Value    Pr > F

    Mean       1     97.9520622     97.9520622       1.50    0.2672
    Error      6    393.0310559     65.5051760

Contrast Variable: strength_3

    Source    DF    Type III SS    Mean Square    F Value    Pr > F

    Mean       1     66.7016645     66.7016645       3.78    0.1000
    Error      6    106.0000000     17.6666667
```

```
                         The GLM Procedure
                Repeated Measures Analysis of Variance
               Analysis of Variance of Contrast Variables

eye_N represents the contrast between the nth level of eye and the last
strength_N represents the nth degree polynomial contrast for strength

Contrast Variable: eye_1*strength_1

     Source    DF     Type III SS    Mean Square    F Value    Pr > F

     Mean       1      1.00621118     1.00621118       0.08    0.7857
     Error      6     74.64596273    12.44099379

Contrast Variable: eye_1*strength_2

     Source    DF     Type III SS    Mean Square    F Value    Pr > F

     Mean       1      56.0809939     56.0809939       1.27    0.3029
     Error      6     265.0789321     44.1798220

Contrast Variable: eye_1*strength_3

     Source    DF     Type III SS    Mean Square    F Value    Pr > F

     Mean       1      24.1627950     24.1627950       1.19    0.3180
     Error      6     122.2751052     20.3791842
```

Display 7.3

Exercises

7.1 Plot the left and right eye means for the different lens strengths. Include standard error bias on the plot.

7.2 Examine the raw data graphically in some way to assess whether there is any evidence of outliers. If there is repeat the analyses described in the text.

7.3 Find the correlations between the repeated measures for the data used in this chapter. Does the pattern of the observed correlations lead to an explanation for the different results produced by the univariate and multivariate treatment of these data?

Chapter 8

Logistic Regression: Psychiatric Screening, Plasma Proteins, and Danish Do-It-Yourself

8.1 Description of Data

This chapter examines three data sets. The first, shown in Display 8.1, arises from a study of a psychiatric screening questionnaire called the GHQ (General Health Questionnaire; see Goldberg [1972]). Here, the question of interest is how "caseness" is related to gender and GHQ score.

The second data set, shown in Display 8.2, was collected to examine the extent to which erythrocyte sedimentation rate (ESR) (i.e., the rate at which red blood cells [erythocytes] settle out of suspension in blood plasma) is related to two plasma proteins: fibrinogen and γ-globulin, both measured in gm/l. The ESR for a "healthy" individual should be less than 20 mm/h and, because the absolute value of ESR is relatively unimportant, the response variable used here denotes whether or not this is the case. A response of zero signifies a healthy individual (ESR < 20), while a response of unity refers to an unhealthy individual (ESR ≥ 20). The aim of the analysis for these data is to determine the strength of any relationship between the ESR level and the levels of the two plasmas.

GHQ Score	*Sex*	*Number of Cases*	*Number of Non-cases*
0	F	4	80
1	F	4	29
2	F	8	15
3	F	6	3
4	F	4	2
5	F	6	1
6	F	3	1
7	F	2	0
8	F	3	0
9	F	2	0
10	F	1	0
0	M	1	36
1	M	2	25
2	M	2	8
3	M	1	4
4	M	3	1
5	M	3	1
6	M	2	1
7	M	4	2
8	M	3	1
9	M	2	0
10	M	2	0

Note: F: Female, M: Male.

Display 8.1

The third data set is given in Display 8.3 and results from asking a sample of employed men, ages 18 to 67, whether, in the preceding year, they had carried out work in their home that they would have previously employed a craftsman to do. The response variable here is the answer (yes/no) to that question. In this situation, we would like to model the relationship between the response variable and four categorical explanatory variables: work, tenure, accommodation type, and age.

Fibrinogen	*γ-Globulin*	*ESR*
2.52	38	0
2.56	31	0
2.19	33	0
2.18	31	0
3.41	37	0
2.46	36	0
3.22	38	0
2.21	37	0
3.15	39	0
2.60	41	0
2.29	36	0
2.35	29	0
5.06	37	1
3.34	32	1
2.38	37	1
3.15	36	0
3.53	46	1
2.68	34	0
2.60	38	0
2.23	37	0
2.88	30	0
2.65	46	0
2.09	44	1
2.28	36	0
2.67	39	0
2.29	31	0
2.15	31	0
2.54	28	0
3.93	32	1
3.34	30	0
2.99	36	0
3.32	35	0

Display 8.2

			Accommodation Type and Age Groups					
			Apartment			House		
Work	*Tenure*	*Response*	*<30*	*31–45*	*46+*	*<30*	*31–45*	*46+*
Skilled	Rent	Yes	18	15	6	34	10	2
		No	15	13	9	28	4	6
	Own	Yes	5	3	1	56	56	35
		No	1	1	1	12	21	8
Unskilled	Rent	Yes	17	10	15	29	3	7
		No	34	17	19	44	13	16
	Own	Yes	2	0	3	23	52	49
		No	3	2	0	9	31	51
Office	Rent	Yes	30	23	21	22	13	11
		No	25	19	40	25	16	12
	Own	Yes	8	5	1	54	191	102
		No	4	2	2	19	76	61

Display 8.3

8.2 The Logistic Regression Model

In linear regression (see Chapter 3), the expected value of a response variable y is modelled as a linear function of the explanatory variables:

$$E(y) = \beta_0 + \beta_1 x_1 + \beta_2 x_2 + \cdots + \beta_p x_p \quad (8.1)$$

For a dichotomous response variable coded 0 and 1, the expected value is simply the probability π that the variable takes the value 1. This could be modelled as in Eq. (8.1), but there are two problems with using linear regression when the response variable is dichotomous:

1. The predicted probability must satisfy $0 \leq \pi \leq 1$, whereas a linear predictor can yield any value from minus infinity to plus infinity.
2. The observed values of y do not follow a normal distribution with mean π, but rather a Bernoulli (or binomial [1, π]) distribution.

In logistic regression, the first problem is addressed by replacing the probability $\pi = E(y)$ on the left-hand side of Eq. (8.1) with the logit transformation of the probability, $\log \pi/(1 - \pi)$. The model now becomes:

$$\text{logit}(\pi) = \log\frac{\pi}{1-\pi} = \beta_0 + \beta_1 x_1 + \cdots + \beta_p x_p \tag{8.2}$$

The logit of the probability is simply the log of the odds of the event of interest. Setting $\boldsymbol{\beta}' = [\beta_0, \beta_1, \cdots, \beta_p]$ and the augmented vector of scores for the *i*th individual as $\boldsymbol{x}_i' = [1, x_{i1}, x_{i2}, \cdots, x_{ip}]$, the predicted probabilities as a function of the linear predictor are:

$$\pi(\boldsymbol{\beta}'\boldsymbol{x}_i) = \frac{\exp(\boldsymbol{\beta}'\boldsymbol{x}_i)}{1 + \exp(\boldsymbol{\beta}'\boldsymbol{x}_i)} \tag{8.3}$$

Whereas the logit can take on any real value, this probability always satisfies $0 \le \pi(\boldsymbol{\beta}'\boldsymbol{x}_i) \le 1$. In a logistic regression model, the parameter β_i associated with explanatory variable x_i is such that $\exp(\beta_i)$ is the odds that $y = 1$ when x_i increases by 1, conditional on the other explanatory variables remaining the same.

Maximum likelihood is used to estimate the parameters of Eq. (8.2), the log-likelihood function being:

$$l(\beta; \boldsymbol{y}) = \sum_i y_i \log[\pi(\boldsymbol{\beta}'\boldsymbol{x}_i)] + (1 - y_i)\log[1 - \pi(\boldsymbol{\beta}'\boldsymbol{x}_i)] \tag{8.4}$$

where $\boldsymbol{y}' = [y_1, y_2, \cdots, y_n]$ are the n observed values of the dichotomous response variable. This log-likelihood is maximized numerically using an iterative algorithm. For full details of logistic regression, see, for example, Collett (1991).

8.3 Analysis Using SAS

8.3.1 GHQ Data

Assuming the data are in the file 'ghq.dat' in the current directory and that the data values are separated by tabs, they can be read in as follows:

```
data ghq;
   infile 'ghq.dat' expandtabs;
   input ghq sex $ cases noncases;
   total=cases+noncases;
   prcase=cases/total;
run;
```

The variable **prcase** contains the observed probability of being a case. This can be plotted against **ghq** score as follows:

```
proc gplot data=ghq;
   plot prcase*ghq;
run;
```

The resulting plot is shown in Display 8.4. Clearly, as the GHQ score increases, the probability of being considered a case increases.

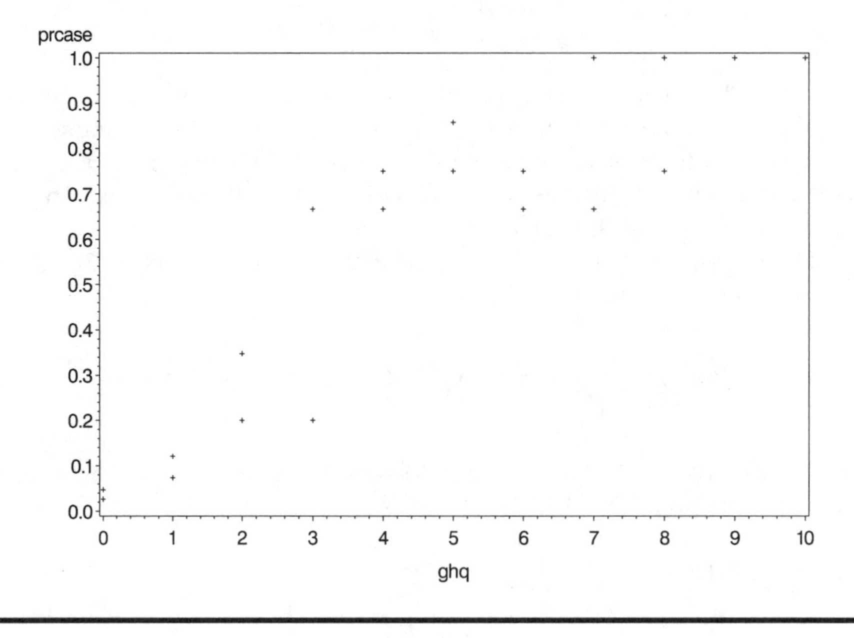

Display 8.4

It is a useful exercise to compare the results of fitting both a simple linear regression and a logistic regression to these data using the single explanatory variable GHQ score. First we perform a linear regression using **proc reg**:

```
proc reg data=ghq;
   model prcase=ghq;
   output out=rout p=rpred;
run;
```

The **output** statement creates an output data set that contains all the original variables plus those created by options. The **p=rpred** option specifies that the predicted values are included in a variable named rpred. The out=rout option specifies the name of the data set to be created.

We then calculate the predicted values from a logistic regression, using proc logistic, in the same way:

```
proc logistic data=ghq;
   model cases/total=ghq;
   output out=lout p=lpred;
run;
```

There are two forms of model statement within proc logistic. This example shows the events/trials syntax, where two variables are specified separated by a slash. The alternative is to specify a single binary response variable before the equal sign.

The two output data sets are combined in a short data step. Because proc gplot plots the data in the order in which they occur, if the points are to be joined by lines it may be necessary to sort the data set into the appropriate order. Both sets of predicted probabilities are to be plotted on the same graph (Display 8.5), together with the observed values; thus, three symbol statements are defined to distinguish them:

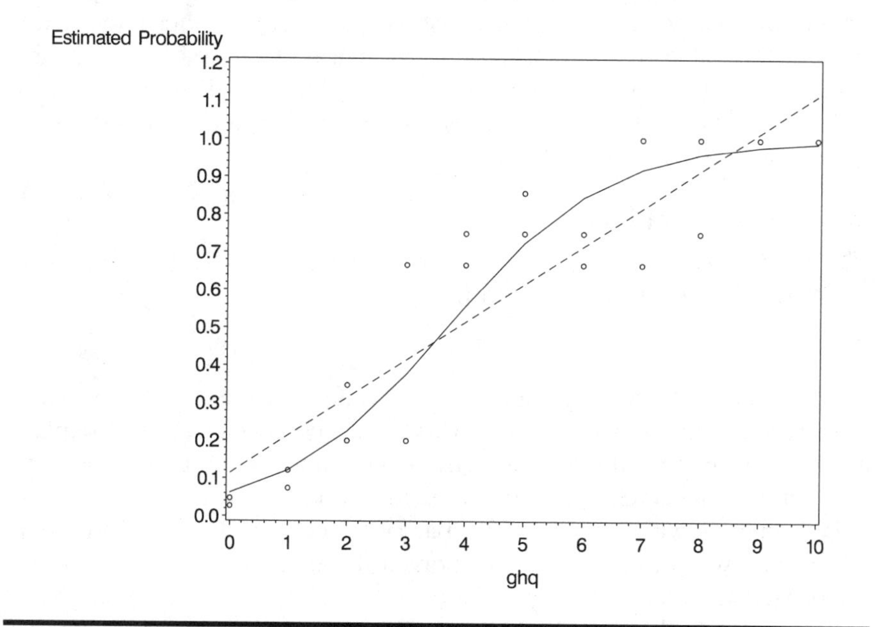

Display 8.5

```
data lrout;
   set rout;
   set lout;

proc sort data=lrout;
   by ghq;

symbol1 i=join v=none l=1;
symbol2 i=join v=none l=2;
symbol3 v=circle;

proc gplot data=lrout;
   plot (rpred lpred prcase)*ghq /overlay;
run;
```

The problems of using the unsuitable linear regression model become apparent on studying Display 8.5. Using this model, two of the predicted values are greater than 1, but the response is a probability constrained to be in the interval (0,1). Additionally, the model provides a very poor fit for the observed data. Using the logistic model, on the other hand, leads to predicted values that are satisfactory in that they all lie between 0 and 1, and the model clearly provides a better description of the observed data.

Next we extend the logistic regression model to include both ghq score and sex as explanatory variables:

```
proc logistic data=ghq;
   class sex;
   model cases/total=sex ghq;
run;
```

The class statement specifies classification variables, or factors, and these can be numeric or character variables. The specification of explanatory effects in the model statement is the same as for proc glm:, with main effects specified by variable names and interactions by joining variable names with asterisks. The bar operator can also be used as an abbreviated way of entering interactions if these are to be included in the model (see Chapter 5).

The output is shown in Display 8.6. The results show that the estimated parameters for both sex and GHQ are significant beyond the 5% level. The parameter estimates are best interpreted if they are converted into odds ratios by exponentiating them. For GHQ, for example, this leads to an odds ratio estimate of exp(0.7791) (i.e., 2.180), with a 95% confidence interval of (1.795, 2.646). A unit increase in GHQ increases the odds of being a case between about 1.8 and 3 times, conditional on sex.

The same procedure can be applied to the parameter for sex, but more care is needed here because the Class Level Information in Display 8.5 shows that sex is coded 1 for females and –1 for males. Consequently, the required odds ratio is exp(2×0.468) (i.e., 2.55), with a 95% confidence interval of (1.088, 5.974). Being female rather than male increases the odds of being a case between about 1.1 and 6 times, conditional on GHQ.

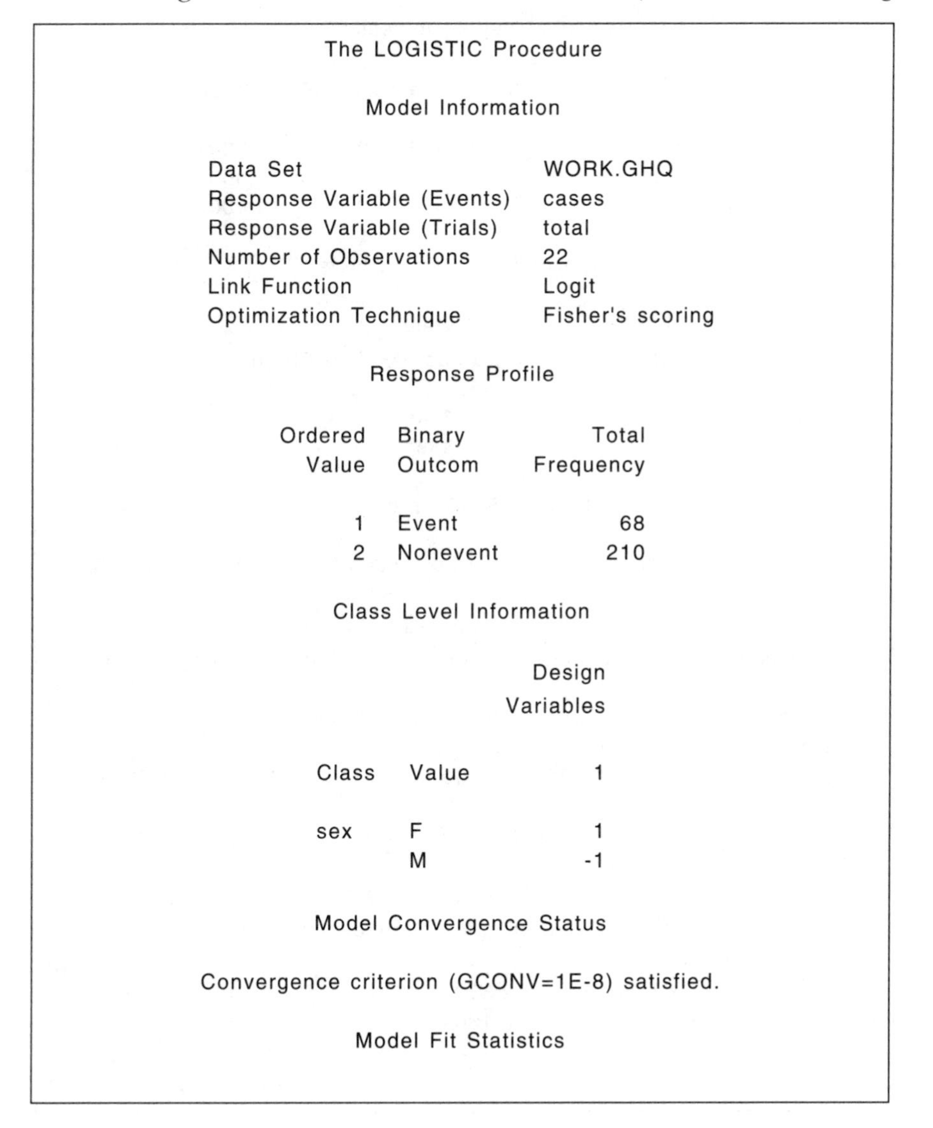

```
                    The LOGISTIC Procedure

                      Model Information

        Data Set                           WORK.GHQ
        Response Variable (Events)         cases
        Response Variable (Trials)         total
        Number of Observations             22
        Link Function                      Logit
        Optimization Technique             Fisher's scoring

                       Response Profile

               Ordered    Binary             Total
                 Value    Outcom         Frequency

                     1    Event                 68
                     2    Nonevent             210

                  Class Level Information

                                        Design
                                      Variables

                   Class    Value             1

                   sex      F                 1
                            M                -1

                  Model Convergence Status

        Convergence criterion (GCONV=1E-8) satisfied.

                     Model Fit Statistics
```

Criterion	Intercept Only	Intercept and Covariates
AIC	311.319	196.126
SC	314.947	207.009
-2 Log L	309.319	190.126

Testing Global Null Hypothesis: BETA=0

Test	Chi-Square	DF	Pr > ChiSq
Likelihood Ratio	119.1929	2	<.0001
Score	120.1327	2	<.0001
Wald	61.9555	2	<.0001

Type III Analysis of Effects

Effect	DF	Wald Chi-Square	Pr > ChiSq
sex	1	4.6446	0.0312
ghq	1	61.8891	<.0001

The LOGISTIC Procedure

Analysis of Maximum Likelihood Estimates

Parameter	DF	Estimate	Standard Error	Chi-Square	Pr > ChiSq
Intercept	1	-2.9615	0.3155	88.1116	<.0001
sex F	1	0.4680	0.2172	4.6446	0.0312
ghq	1	0.7791	0.0990	61.8891	<.0001

Odds Ratio Estimates

Effect	Point Estimate	95% Wald Confidence Limits	
sex F vs M	2.550	1.088	5.974
ghq	2.180	1.795	2.646

```
Association of Predicted Probabilities and Observed Responses

        Percent Concordant      85.8    Somers' D    0.766
        Percent Discordant       9.2    Gamma        0.806
        Percent Tied             5.0    Tau-a        0.284
        Pairs                  14280    c            0.883
```

Display 8.6

8.3.2 ESR and Plasma Levels

We now move on to examine the ESR data in Display 8.2. The data are first read in for analysis using the following SAS code:

```
data plasma;
   infile 'n:\handbook2\datasets\plasma.dat';
   input fibrinogen gamma esr;
run;
```

We can try to identify which of the two plasma proteins — fibrinogen or γ-globulin — has the strongest relationship with ESR level by fitting a logistic regression model and allowing here, backward elimination of variables as described in Chapter 3 for multiple regression, although the elimination criterion is now based on a likelihood ratio statistic rather than an *F*-value.

```
proc logistic data=plasma desc;
   model esr=fibrinogen gamma fibrinogen*gamma / selec-
tion=backward;
run;
```

Where a binary response variable is used on the model statement, as opposed to the events/trials used for the GHQ data, SAS models the lower of the two response categories as the "event." However, it is common practice for a binary response variable to be coded 0,1 with 1 indicating a response (or event) and 0 indicating no response (or a non-event). In this case, the seemingly perverse, default in SAS will be to model the probability of a non-event. The desc (descending) option in the proc statement reverses this behaviour.

It is worth noting that when the model selection option is forward, backward, or stepwise, SAS preserves the hierarchy of effects by default.

For an interaction effect to be allowed in the model, all the lower-order interactions and main effects that it implies must also be included.

The results are given in Display 8.7. We see that both the fibrinogen $\times$ γ-globulin interaction effect and the γ-globulin main effect are eliminated from the initial model. It appears that only fibrinogen level is predictive of ESR level.

```
                    The LOGISTIC Procedure

                      Model Information

          Data Set                        WORK.PLASMA
          Response Variable               esr
          Number of Response Levels       2
          Number of Observations          32
          Link Function                   Logit
          Optimisation Technique          Fisher's scoring

                      Response Profile

                 Ordered                   Total
                   Value     esr       Frequency

                       1       1               6
                       2       0              26

               Backward Elimination Procedure

Step 0. The following effects were entered:

Intercept fibrinogen gamma fibrinogen*gamma

                  Model Convergence Status

          Convergence criterion (GCONV=1E-8) satisfied.

                    Model Fit Statistics

                                          Intercept
                            Intercept         and
             Criterion           Only      Covariates

             AIC               32.885          28.417
             SC                34.351          34.280
             -2 Log L          30.885          20.417
```

```
               Testing Global Null Hypothesis: BETA=0

            Test                 Chi-Square    DF    Pr > ChiSq

            Likelihood Ratio        10.4677     3        0.0150
            Score                    8.8192     3        0.0318
            Wald                     4.7403     3        0.1918

Step 1. Effect fibrinogen*gamma is removed:

                      Model Convergence Status

             Convergence criterion (GCONV=1E-8) satisfied.

                       The LOGISTIC Procedure

                        Model Fit Statistics

                                             Intercept
                              Intercept         and
                 Criterion       Only        Covariates

                 AIC           32.885          28.971
                 SC            34.351          33.368
                 -2 Log L      30.885          22.971

               Testing Global Null Hypothesis: BETA=0

           Test                 Chi-Square    DF    Pr > ChiSq

           Likelihood Ratio         7.9138     2        0.0191
           Score                    8.2067     2        0.0165
           Wald                     4.7561     2        0.0927

                     Residual Chi-Square Test

                   Chi-Square    DF    Pr > ChiSq

                       2.6913     1        0.1009

Step 2. Effect gamma is removed:

                      Model Convergence Status

             Convergence criterion (GCONV=1E-8) satisfied.
```

Model Fit Statistics

Criterion	Intercept Only	Intercept and Covariates
AIC	32.885	28.840
SC	34.351	31.772
-2 Log L	30.885	24.840

Testing Global Null Hypothesis: BETA=0·

Test	Chi-Square	DF	Pr > ChiSq
Likelihood Ratio	6.0446	1	0.0139
Score	6.7522	1	0.0094
Wald	4.1134	1	0.0425

Residual Chi-Square Test

Chi-Square	DF	Pr > ChiSq
4.5421	2	0.1032

NOTE: No (additional) effects met the 0.05 significance level for removal from the model.

The LOGISTIC Procedure

Summary of Backward Elimination

Step	Effect Removed	DF	Number In	Wald Chi-Square	Pr > ChiSq
1	fibrinogen*gamma	1	2	2.2968	0.1296
2	gamma	1	1	1.6982	0.1925

Analysis of Maximum Likelihood Estimates

Parameter	DF	Estimate	Standard Error	Chi-Square	Pr > ChiSq
Intercept	1	-6.8451	2.7703	6.1053	0.0135
fibrinogen	1	1.8271	0.9009	4.1134	0.0425

Odds Ratio Estimates

Effect	Point Estimate	95% Wald Confidence Limits	
fibrinogen	6.216	1.063	36.333

Association of Predicted Probabilities and Observed Responses

Percent Concordant	71.2	Somers' D	0.429
Percent Discordant	28.2	Gamma	0.432
Percent Tied	0.6	Tau-a	0.135
Pairs	156	c	0.715

Display 8.7

It is useful to look at a graphical display of the final model selected and the following code produces a plot of predicted values from the fibrinogen-only logistic model along with the observed values of ESR (remember that these can only take the values of 0 or 1). The plot is shown in Display 8.8.

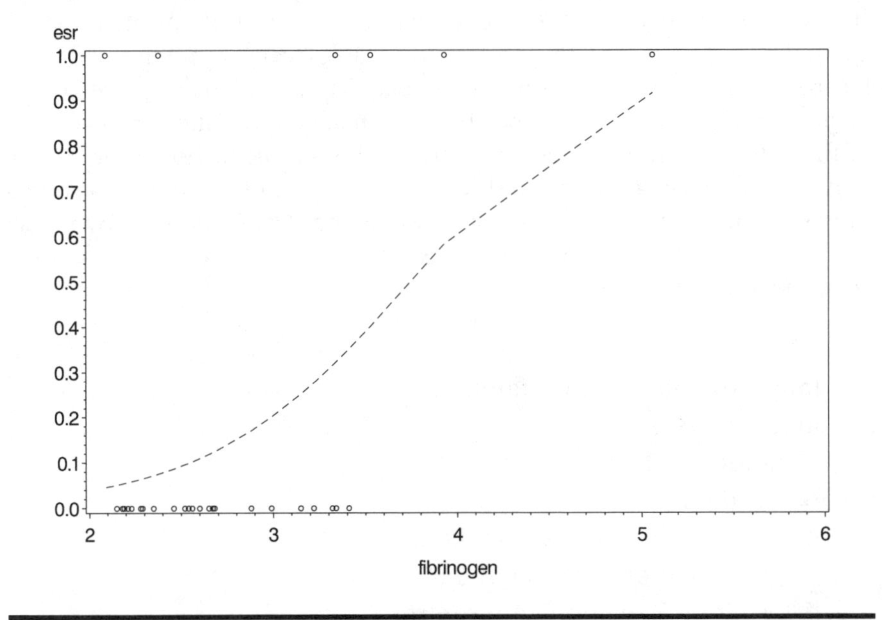

Display 8.8

```
proc logistic data=plasma desc;
   model esr=fibrinogen;
   output out=lout p=lpred;

proc sort data=lout;
   by fibrinogen;

symbol1 i=none v=circle;
symbol2 i=join v=none;
proc gplot data=lout;
   plot (esr lpred)*fibrinogen /overlay ;
run;
```

Clearly, an increase in fibrinogen level is associated with an increase in the probability of the individual being categorised as unhealthy.

8.3.3 Danish Do-It-Yourself

Assuming that the data shown in Display 8.3 are in a file 'diy.dat' in the current directory and the values are separated by tabs, the following data step can be used to create a SAS data set for analysis. As in previous examples, the values of the grouping variables can be determined from the row and column positions in the data. An additional feature of this data set is that each cell of the design contains two data values: counts of those who answered "yes" and "no" to the question about work in the home. Each observation in the data set needs both of these values so that the events/trials syntax can be used in proc logistic. To do this, two rows of data are input at the same time: six counts of "yes" responses and the corresponding "no" responses.

```
data diy;
   infile 'diy.dat' expandtabs;
   input y1-y6 / n1-n6;
      length work $9.;
work='Skilled';
   if _n_ > 2 then work='Unskilled';
   if _n_ > 4 then work='Office';
   if _n_ in(1,3,5) then tenure='rent';
                         else tenure='own';
   array yall {6} y1-y6;
```

```
      array nall {6} n1-n6;
      do i=1 to 6;
         if i>3 then type='house';
                else type='flat';
      agegrp=1;
         if i in(2,5) then agegrp=2;
         if i in(3,6) then agegrp=3;
      yes=yall{i};
         no=nall{i};
         total=yes+no;
         prdiy=yes/total;
         output;
      end;
      drop i y1--n6;
   run;
```

The **expandtabs** option in the **infile** statement allows list input to be used. The **input** statement reads two lines of data from the file. Six data values are read from the first line into variables **y1** to **y6**. The slash that follows tells SAS to go to the next line and six values from there are read into variables **n1** to **n6**.

There are 12 lines of data in the file; but because each pass through the data step is reading a pair of lines, the automatic variable **_n_** will range from 1 to 6. The appropriate values of the work and tenure variables can then be assigned accordingly. Both are character variables and the **length** statement specifies that work is nine characters. Without this, its length would be determined from its first occurrence in the data step. This would be the statement **work='skilled'**; and a length of seven characters would be assigned, insufficient for the value **'unskilled'**.

The variables containing the yes and no responses are declared as arrays and processed in parallel inside a **do** loop. The values of age group and accommodation type are determined from the index of the **do** loop (i.e., from the column in the data). Counts of yes and corresponding no responses are assigned to the variables **yes** and **no**, their sum assigned to **total**, and the observed probability of a yes to **prdiy**. The **output** statement within the **do** loop writes six observations to the data set. (See Chapter 5 for a more complete explanation.)

As usual with a complicated data step such as this, it is wise to check the results; for example, with **proc print**.

A useful starting point in examining these data is a tabulation of the observed probabilities using **proc tabulate**:

```
proc tabulate data=diy order=data f=6.2;
   class work tenure type agegrp;
   var prdiy;
   table work*tenure all,
         (type*agegrp all)*prdiy*mean;
run;
```

Basic use of **proc tabulate** was described in Chapter 5. In this example, the **f=** option specifies a format for the cell entries of the table, namely six columns with two decimal places. It also illustrates the use of the keyword **all** for producing totals. The result is shown in Display 8.9. We see that there are considerable differences in the observed probabilities, suggesting that some, at least, of the explanatory variables may have an effect.

		type						
		flat			house			
		agegrp			agegrp			
		1	2	3	1	2	3	All
		prdiy	prdiy	prdiy	prdiy	prdiy	prdiy	prdiy
		Mean	Mean	Mean	Mean	Mean	Mean	Mean
work	tenure							
Skilled	rent	0.55	0.54	0.40	0.55	0.71	0.25	0.50
	own	0.83	0.75	0.50	0.82	0.73	0.81	0.74
Unskilled	rent	0.33	0.37	0.44	0.40	0.19	0.30	0.34
	own	0.40	0.00	1.00	0.72	0.63	0.49	0.54
Office	rent	0.55	0.55	0.34	0.47	0.45	0.48	0.47
	own	0.67	0.71	0.33	0.74	0.72	0.63	0.63
All		0.55	0.49	0.50	0.62	0.57	0.49	0.54

Display 8.9

We continue our analysis of the data with a backwards elimination logistic regression for the main effects of the four explanatory variables only.

```
proc logistic data=diy;
   class work tenure type agegrp /param=ref ref=first;
   model yes/total=work tenure type agegrp / selection=back
ward;
run;
```

All the predictors are declared as classfication variables, or factors, on the **class** statement. The **param** option specifies reference coding (more commonly referred to as dummy variable coding), with the **ref** option setting the first category to be the reference category. The output is shown in Display 8.10.

```
                 The LOGISTIC Procedure

                   Model Information

      Data Set                       WORK.DIY
      Response Variable (Events)     yes
      Response Variable (Trials)     total
      Number of Observations         36
      Link Function                  Logit
      Optimization Technique         Fisher's scoring

                   Response Profile

           Ordered    Binary            Total
             Value    Outcome       Frequency

                 1    Event               932
                 2    Nonevent            659

            Backward Elimination Procedure

                Class Level Information

                                       Design
                                     Variables

            Class      Value          1      2

            work       Office         0      0
                       Skilled        1      0
                       Unskilled      0      1

            tenure     own            0
                       rent           1

            type       flat           0
                       house          1

            agegrp     1              0      0
                       2              1      0
                       3              0      1
```

```
Step 0. The following effects were entered:

Intercept work tenure type agegrp

                    Model Convergence Status

          Convergence criterion (GCONV=1E-8) satisfied.

                      Model Fit Statistics

                                     Intercept
                        Intercept       and
            Criterion     Only      Covariates

            AIC         2160.518     2043.305
            SC          2165.890     2080.910
            -2 Log L    2158.518     2029.305

                    The LOGISTIC Procedure

            Testing Global Null Hypothesis: BETA=0

        Test                 Chi-Square    DF    Pr > ChiSq

        Likelihood Ratio       129.2125     6        <.0001
        Score                  126.8389     6        <.0001
        Wald                   119.6073     6        <.0001

Step 1. Effect type is removed:

                    Model Convergence Status

          Convergence criterion (GCONV=1E-8) satisfied.

                      Model Fit Statistics

                                     Intercept
                        Intercept       and
            Criterion     Only      Covariates

            AIC         2160.518     2041.305
            SC          2165.890     2073.538
            -2 Log L    2158.518     2029.305
```

Testing Global Null Hypothesis: BETA=0

Test	Chi-Square	DF	Pr > ChiSq
Likelihood Ratio	129.2122	5	<.0001
Score	126.8382	5	<.0001
Wald	119.6087	5	<.0001

Residual Chi-Square Test

Chi-Square	DF	Pr > ChiSq
0.0003	1	0.9865

NOTE: No (additional) effects met the 0.05 significance level for removal from the model.

Summary of Backward Elimination

Step	Effect Removed	DF	Number In	Wald Chi-Square	Pr > ChiSq
1	type	1	3	0.0003	0.9865

Type III Analysis of Effects

Effect	DF	Wald Chi-Square	Pr > ChiSq
work	2	27.0088	<.0001
tenure	1	78.6133	<.0001
agegrp	2	10.9072	0.0043

The LOGISTIC Procedure

Analysis of Maximum Likelihood Estimates

Parameter		DF	Estimate	Standard Error	Chi-Square	Pr > ChiSq
Intercept		1	1.0139	0.1361	55.4872	<.0001
work	Skilled	1	0.3053	0.1408	4.7023	0.0301
work	Unskilled	1	-0.4574	0.1248	13.4377	0.0002
tenure	rent	1	-1.0144	0.1144	78.6133	<.0001
agegrp	2	1	-0.1129	0.1367	0.6824	0.4088
agegrp	3	1	-0.4364	0.1401	9.7087	0.0018

Odds Ratio Estimates

Effect	Point Estimate	95% Wald Confidence Limits	
work Skilled vs Office	1.357	1.030	1.788
work Unskilled vs Office	0.633	0.496	0.808
tenure rent vs own	0.363	0.290	0.454
agegrp 2 vs 1	0.893	0.683	1.168
agegrp 3 vs 1	0.646	0.491	0.851

Association of Predicted Probabilities and Observed Responses

Percent Concordant	62.8	Somers' D	0.327
Percent Discordant	30.1	Gamma	0.352
Percent Tied	7.1	Tau-a	0.159
Pairs	614188	c	0.663

Display 8.10

Work, tenure, and age group are all selected in the final model; only the type of accommodation is dropped. The estimated conditional odds ratio suggests that skilled workers are more likely to respond "yes" to the question asked than office workers (estimated odds ratio 1.357 with 95% confidence interval 1.030, 1.788). And unskilled workers are less likely than office workers to respond "yes" (0.633, 0.496–0.808). People who rent their home are far less likely to answer "yes" than people who own their home (0.363, 0.290–0.454). Finally, it appears that people in the two younger age groups are more likely to respond "yes" than the oldest respondents.

Exercises

8.1 In the text, a main-effects-only logistic regression was fitted to the GHQ data. This assumes that the effect of GHQ on caseness is the same for men and women. Fit a model where this assumption is not made, and assess which model best fits the data.

8.2 For the ESR and plasma protein data, fit a logistic model that includes quadratic effects for both fibrinogen and γ-globulin. Does the model fit better than the model selected in the text?

8.3 Investigate using logistic regression on the Danish do-it-yourself data allowing for interactions among some factors.

Chapter 9

Generalised Linear Models: School Attendance Amongst Australian School Children

9.1 Description of Data

This chapter reanalyses a number of data sets from previous chapters, in particular the data on school attendance in Australian school children used in Chapter 6. The aim of this chapter is to introduce the concept of generalized linear models and to illustrate how they can be applied in SAS using proc genmod.

9.2 Generalised Linear Models

The analysis of variance models considered in Chapters 5 and 6 and the multiple regression model described in Chapter 4 are, essentially, completely equivalent. Both involve a linear combination of a set of explanatory variables (dummy variables in the case of analysis of variance) as a model for an observed response variable. And both include residual terms assumed to have a normal distribution. (The equivalence of analysis of variance and multiple regression models is spelt out in more detail in Everitt [2001].)

The logistic regression model encountered in Chapter 8 also has similarities to the analysis of variance and multiple regression models. Again, a linear combination of explanatory variables is involved, although here the binary response variable is not modelled directly (for the reasons outlined in Chapter 8), but via a logistic transformation.

Multiple regression, analysis of variance, and logistic regression models can all, in fact, be included in a more general class of models known as *generalised linear models*. The essence of this class of models is a linear predictor of the form:

$$\eta = \beta_0 + \beta_1 x_1 + \cdots + \beta_p x_p = \boldsymbol{\beta}' \boldsymbol{x} \tag{9.1}$$

where $\boldsymbol{\beta}' = [\beta_0, \beta_1, \cdots, \beta_p]$ and $\boldsymbol{x}' = [1, x_1, x_2, \cdots, x_p]$. The linear predictor determines the expectation, μ, of the response variable. In linear regression, where the response is continuous, μ is directly equated with the linear predictor. This is not sensible when the response is dichotomous because in this case the expectation is a probability that must satisfy $0 \leq \mu \leq 1$. Consequently, in logistic regression, the linear predictor is equated with the logistic function of μ, $\log \mu/(1 - \mu)$.

In the generalised linear model formulation, the linear predictor can be equated with a chosen function of μ, $g(\mu)$, and the model now becomes:

$$\eta = g(\mu) \tag{9.2}$$

The function g is referred to as a *link function*.

In linear regression (and analysis of variance), the probability distribution of the response variable is assumed to be normal with mean μ. In logistic regression, a binomial distribution is assumed with probability parameter μ. Both distributions, the normal and binomial distributions, come from the same family of distributions, called the *exponential family*, and are given by:

$$f(y; \theta, \phi) = \exp\{(y\theta - b(\theta))/a(\phi) + c(y, \phi)\} \tag{9.3}$$

For example, for the normal distribution,

$$f(y;\theta,\phi) = \frac{1}{\sqrt{(2\pi\sigma^2)}}\exp\{-(y-\mu)^2/2\sigma^2\}$$

$$= \exp\left\{(y\mu-\mu^2/2)/\sigma^2 - \frac{1}{2}(y^2/\sigma^2 + \log(2\pi\sigma^2))\right\} \quad (9.4)$$

so that $\theta = \mu$, $b(\theta) = \theta^2/2$, $\phi = \sigma^2$ and $a(\phi) = \phi$.

The parameter θ, a function of μ, is called the *canonical link*. The canonical link is frequently chosen as the link function, although the canonical link is not necessarily more appropriate than any other link. Display 9.1 lists some of the most common distributions and their canonical link functions used in generalised linear models.

Distribution	*Variance Function*	*Dispersion Parameter*	*Link Function*	*$g(\mu) = \theta(\mu)$*
Normal	1	σ^2	Identity	μ
Binomial	$\mu(1-\mu)$	1	Logit	$\log(\mu/(1-\mu))$
Poisson	μ	1	Log	$\log(\mu)$
Gamma	μ^2	v^{-1}	Reciprocal	$1/\mu$

Display 9.1

The mean and variance of a random variable Y having the distribution in Eq. (9.3) are given, respectively, by:

$$E(Y) = b'(0) = \mu \quad (9.5)$$

and

$$\text{var}(Y) = b''(\theta)\, a(\phi) = V(\mu)\, a(\phi) \quad (9.6)$$

where $b'(\theta)$ and $b''(\theta)$ denote the first and second derivative of $b(\theta)$ with respect to θ, and the variance function $V(\mu)$ is obtained by expression $b''(\theta)$ as a function of μ. It can be seen from Eq. (9.4) that the variance for the normal distribution is simply σ^2, regardless of the value of the mean μ, that is, the variance function is 1.

The data on Australian school children will be analysed by assuming a Poisson distribution for the number of days absent from school. The Poisson distribution is the appropriate distribution of the number of events observed if these events occur independently in continuous time at a constant instantaneous probability rate (or incidence rate); see, for example, Clayton and Hills (1993). The Poisson distribution is given by:

$$f(y;\mu) = \mu^y e^{-\mu}/y!, \; y = 0, 1, 2, \cdots \tag{9.7}$$

Taking the logarithm and summing over observations $y_1, y_2, \cdots, y_n$, the log likelihood is

$$l(\mu; y_1, y_2, \cdots, y_n) = \sum_i \{(y_i \ln \mu_i - \mu_i) - \ln(y_i!)\} \tag{9.8}$$

Where $\mu' = [\mu_1 \cdots \mu_n]$ gives the expected values of each observation. Here $\theta = \ln \mu$, $b(\theta) = \exp(\theta)$, $\phi = 1$, and $\text{var}(y) = \exp(\theta) = \mu$. Therefore, the variance of the Poisson distribution is not constant, but equal to the mean. Unlike the normal distribution, the Poisson distribution has no separate parameter for the variance and the same is true of the Binomial distribution. Display 9.1 shows the variance functions and dispersion parameters for some commonly used probability distributions.

9.2.1 Model Selection and Measure of Fit

Lack of fit in a generalized linear model can be expressed by the deviance, which is minus twice the difference between the maximized log-likelihood of the model and the maximum likelihood achievable, that is, the maximized likelihood of the *full* or saturated model. For the normal distribution, the deviance is simply the residual sum of squares. Another measure of lack of fit is the generalized Pearson X^2,

$$X^2 = \sum_i (y_i - \hat{\mu}_i)^2 / V(\hat{\mu}_i) \tag{9.9}$$

which, for the Poisson distribution, is just the familiar statistic for two-way cross-tabulations (since $V(\hat{\mu}) = \hat{\mu}$). Both the deviance and Pearson X^2 have chi-square distributions when the sample size tends to infinity. When the dispersion parameter ϕ is fixed (not estimated), an analysis of variance can be used for testing nested models in the same way as analysis of variance is used for linear models. The difference in deviance between two models is simply compared with the chi-square distribution, with degrees of freedom equal to the difference in model degrees of freedom.

The Pearson and deviance residuals are defined as the (signed) square roots of the contributions of the individual observations to the Pearson

X^2 and deviance, respectively. These residuals can be used to assess the appropriateness of the link and variance functions.

A relatively common phenomenon with binary and count data is *overdispersion*, that is, the variance is greater than that of the assumed distribution (binomial and Poisson, respectively). This overdispersion may be due to extra variability in the parameter μ, which has not been completely explained by the covariates. One way of addressing the problem is to allow μ to vary randomly according to some (prior) distribution and to assume that conditional on the parameter having a certain value, the response variable follows the binomial (or Poisson) distribution. Such models are called *random effects models* (see Pinheiro and Bates [2000] and Chapter 11).

A more pragmatic way of accommodating overdispersion in the model is to assume that the variance is proportional to the variance function, but to estimate the dispersion rather than assuming the value 1 appropriate for the distributions. For the Poisson distribution, the variance is modelled as:

$$\text{var}(Y) = \phi\mu \qquad (9.10)$$

where ϕ is the estimated from the deviance or Pearson X^2. (This is analogous to the estimation of the residual variance in linear regression models from the residual sums of squares.) This parameter is then used to scale the estimated standard errors of the regression coefficients. This approach of assuming a variance function that does not correspond to any probability distribution is an example of *quasi-likelihood*. See McCullagh and Nelder (1989) for more details on generalised linear models.

9.3 Analysis Using SAS

Within SAS, the **genmod** procedure uses the framework described in the previous section to fit generalised linear models. The distributions covered include those shown in Display 9.1, plus the inverse Gaussian, negative binomial, and multinomial.

To first illustrate the use of **proc genmod**, we begin by replicating the analysis of U.S. crime rates presented in Chapter 4 using the subset of explanatory variables selected by stepwise regression. We assume the data have been read into a SAS data set **uscrime** as described there.

```
proc genmod data=uscrime;
   model R=ex1 x ed age u2 / dist=normal link=identity;
run;
```

The **model** statement specifies the regression equation in much the same way as for **proc glm** described in Chapter 6. For a binomial response, the events/trials syntax described in Chapter 8 for **proc logistic** can also be used. The distribution and link function are specified as options in the **model** statement. Normal and identity can be abbreviated to N and id, respectively. The output is shown in Display 9.2. The parameter estimates are equal to those obtained in Chapter 4 using **proc reg** (see Display 4.5), although the standard errors are not identical. The deviance value of 495.3383 is equal to the error mean square in Display 4.5.

```
                          The GENMOD Procedure

                           Model Information

                Data Set                      WORK.USCRIME
                Distribution                        Normal
                Link Function                     Identity
                Dependent Variable                       R
                Observations Used                       47

                Criteria For Assessing Goodness-Of-Fit

          Criterion                 DF          Value       Value/DF

          Deviance                  41     20308.8707       495.3383
          Scaled Deviance           41        47.0000         1.1463
          Pearson Chi-Square        41     20308.8707       495.3383
          Scaled Pearson X2         41        47.0000         1.1463
          Log Likelihood                    -209.3037

Algorithm converged.

                     Analysis Of Parameter Estimates

                            Standard   Wald 95% Confidence      Chi-
Parameter   DF   Estimate      Error          Limits          Square   Pr > ChiSq

Intercept    1   -528.856    93.0407   -711.212   -346.499     32.31       <.0001
Ex1          1     1.2973     0.1492     1.0050     1.5897     75.65       <.0001
X            1     0.6463     0.1443     0.3635     0.9292     20.06       <.0001
Ed           1     2.0363     0.4628     1.1294     2.9433     19.36       <.0001
Age          1     1.0184     0.3447     0.3428     1.6940      8.73       0.0031
U2           1     0.9901     0.4223     0.1625     1.8178      5.50       0.0190
Scale        1    20.7871     2.1440    16.9824    25.4442

NOTE: The scale parameter was estimated by maximum likelihood.
```

Display 9.2

Now we can move on to a more interesting application of generalized linear models involving the data on Australian children's school attendance, used previously in Chapter 6 (see Display 6.1). Here, because the response variable — number of days absent — is a count, we will use a Poisson distribution and a log link.

Assuming that the data on Australian school attendance have been read into a SAS data set, **ozkids**, as described in Chapter 6, we fit a main effects model as follows.

```
proc genmod data=ozkids;
   class origin sex grade type;
   model days=sex origin type grade / dist=p link=log type1
type3;
run;
```

The predictors are all categorical variables and thus must be declared as such with a **class** statement. The Poisson probability distribution with a log link are requested with Type 1 and Type 3 analyses. These are analogous to Type I and Type III sums of squares discussed in Chapter 6. The results are shown in Display 9.3. Looking first at the LR statistics for each of the main effects, we see that both Type 1 and Type 3 analyses lead to very similar conclusions, namely that each main effect is significant. For the moment, we will ignore the Analysis of Parameter Estimates part of the output and examine instead the criteria for assessing goodness-of-fit. In the absence of overdispersion, the dispersion parameters based on the Pearson X^2 of the deviance should be close to 1. The values of 13.6673 and 12.2147 given in Display 9.3 suggest, therefore, that there is overdispersion; and as a consequence, the P-values in this display may be too low.

```
                    The GENMOD Procedure

                     Model Information

         Data Set                     WORK.OZKIDS
         Distribution                     Poisson
         Link Function                        Log
         Dependent Variable                  days
         Observations Used                    154
```

Class Level Information

Class	Levels	Values
origin	2	A N
sex	2	F M
grade	4	F0 F1 F2 F3
type	2	AL SL

Criteria For Assessing Goodness-Of-Fit

Criterion	DF	Value	Value/DF
Deviance	147	1795.5665	12.2147
Scaled Deviance	147	1795.5665	12.2147
Pearson Chi-Square	147	2009.0882	13.6673
Scaled Pearson X2	147	2009.0882	13.6673
Log Likelihood		4581.1746	

Algorithm converged.

Analysis Of Parameter Estimates

Parameter		DF	Estimate	Standard Error	Wald 95% Confidence Limits		Chi-Square	Pr > ChiSq
Intercept		1	2.7742	0.0628	2.6510	2.8973	1949.78	<.0001
sex	F	1	-0.1405	0.0416	-0.2220	-0.0589	11.39	0.0007
sex	M	0	0.0000	0.0000	0.0000	0.0000	.	.
origin	A	1	0.4951	0.0412	0.4143	0.5758	144.40	<.0001
origin	N	0	0.0000	0.0000	0.0000	0.0000	.	.
type	AL	1	-0.1300	0.0442	-0.2166	-0.0435	8.67	0.0032
type	SL	0	0.0000	0.0000	0.0000	0.0000	.	.
grade	F0	1	-0.2060	0.0629	-0.3293	-0.0826	10.71	0.0011
grade	F1	1	-0.4718	0.0614	-0.5920	-0.3515	59.12	<.0001
grade	F2	1	0.1108	0.0536	0.0057	0.2158	4.27	0.0387
grade	F3	0	0.0000	0.0000	0.0000	0.0000	.	.
Scale		0	1.0000	0.0000	1.0000	1.0000		

NOTE: The scale parameter was held fixed.

LR Statistics For Type 1 Analysis

Source	Deviance	DF	Chi-Square	Pr > ChiSq
Intercept	2105.9714			
sex	2086.9571	1	19.01	<.0001
origin	1920.3673	1	166.59	<.0001
type	1917.0156	1	3.35	0.0671
grade	1795.5665	3	121.45	<.0001

```
The GENMOD Procedure

LR Statistics For Type 3 Analysis

                        Chi-
Source      DF        Square     Pr > ChiSq

sex          1         11.38         0.0007
origin       1        148.20         <.0001
type         1          8.65         0.0033
grade        3        121.45         <.0001
```

Display 9.3

To rerun the analysis allowing for overdispersion, we need an estimate of the dispersion parameter ϕ. One strategy is to fit a model that contains a sufficient number of parameters so that all systematic variation is removed, estimate ϕ from this model as the deviance of Pearson X^2 divided by its degrees of freedom, and then use this estimate in fitting the required model.

Thus, here we first fit a model with all first-order interactions included, simply to get an estimate of ϕ. The necessary SAS code is

```
proc genmod data=ozkids;
   class origin sex grade type;
   model days=sex|origin|type|grade@2 / dist=p link=log
scale=d;
run;
```

The scale=d option in the model statement specifies that the scale parameter is to be estimated from the deviance. The model statement also illustrates a modified use of the bar operator. By appending @2, we limit its expansion to terms involving two effects. This leads to an estimate of ϕ of 3.1892.

We now fit a main effects model allowing for overdispersion by specifying scale =3.1892 as an option in the model statement.

```
proc genmod data=ozkids;
   class origin sex grade type;
   model days=sex origin type grade / dist=p link=log type1
type3 scale=3.1892;
output out=genout pred=pr_days stdreschi=resid;
run;
```

The **output** statement specifies that the predicted values and standardized Pearson (Chi) residuals are to be saved in the variables **pr_days** and **resid**, respectively, in the data set **genout**.

The new results are shown in Display 9.4. Allowing for overdispersion has had no effect on the regression coefficients, but a large effect on the P-values and confidence intervals so that sex and type are no longer significant. Interpretation of the significant effect of, for example, origin is made in terms of the logs of the predicted mean counts. Here, the estimated coefficient for origin 0.4951 indicates that the log of the predicted mean number of days absent from school for Aboriginal children is 0.4951 higher than for white children, conditional on the other variables. Exponentiating the coefficient yield count ratios, that is, 1.64 with corresponding 95% confidence interval (1.27, 2.12). Aboriginal children have between about one and a quarter to twice as many days absent as white children.

The standardized residuals can be plotted against the predicted values using **proc gplot**.

```
proc gplot data=genout;
   plot resid*pr_days;
run;
```

The result is shown in Display 9.5. This plot does not appear to give any cause for concern.

```
                    The GENMOD Procedure

                     Model Information

          Data Set                      WORK.OZKIDS
          Distribution                      Poisson
          Link Function                         Log
          Dependent Variable                   days
          Observations Used                     154

                  Class Level Information

             Class     Levels    Values

             origin       2      A N
             sex          2      F M
             grade        4      F0 F1 F2 F3
             type         2      AL SL
```

Criteria For Assessing Goodness Of Fit

Criterion	DF	Value	Value/DF
Deviance	147	1795.5665	12.2147
Scaled Deviance	147	176.5379	1.2009
Pearson Chi-Square	147	2009.0882	13.6673
Scaled Pearson X2	147	197.5311	1.3437
Log Likelihood		450.4155	

Algorithm converged.

Analysis Of Parameter Estimates

Parameter		DF	Estimate	Standard Error	Wald 95% Confidence Limits		Chi-Square	Pr > ChiSq
Intercept		1	2.7742	0.2004	2.3815	3.1669	191.70	<.0001
sex	F	1	-0.1405	0.1327	-0.4006	0.1196	1.12	0.2899
sex	M	0	0.0000	0.0000	0.0000	0.0000	.	.
origin	A	1	0.4951	0.1314	0.2376	0.7526	14.20	0.0002
origin	N	0	0.0000	0.0000	0.0000	0.0000	.	.
type	AL	1	-0.1300	0.1408	-0.4061	0.1460	0.85	0.3559
type	SL	0	0.0000	0.0000	0.0000	0.0000	.	.
grade	F0	1	-0.2060	0.2007	-0.5994	0.1874	1.05	0.3048
grade	F1	1	-0.4718	0.1957	-0.8553	-0.0882	5.81	0.0159
grade	F2	1	0.1108	0.1709	-0.2242	0.4457	0.42	0.5169
grade	F3	0	0.0000	0.0000	0.0000	0.0000	.	.
Scale		0	3.1892	0.0000	3.1892	3.1892		

NOTE: The scale parameter was held fixed.

LR Statistics For Type 1 Analysis

Source	Deviance	DF	Chi-Square	Pr > ChiSq
Intercept	2105.9714			
sex	2086.9571	1	1.87	0.1715
origin	1920.3673	1	16.38	<.0001
type	1917.0156	1	0.33	0.5659
grade	1795.5665	3	11.94	0.0076

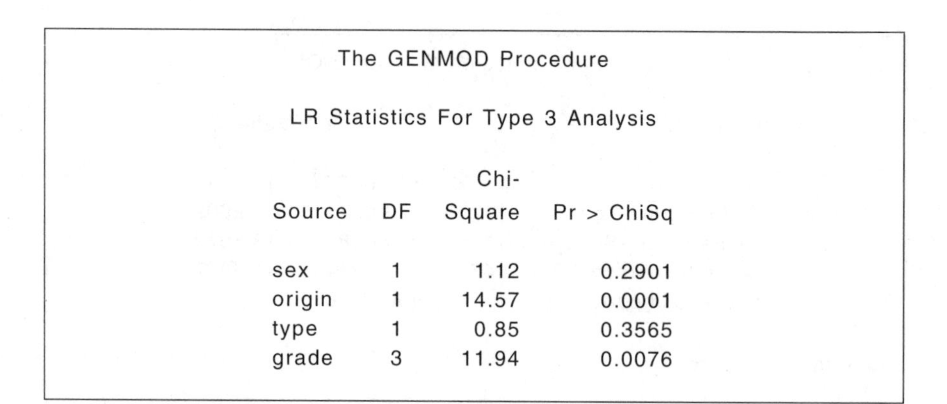

The GENMOD Procedure

LR Statistics For Type 3 Analysis

Source	DF	Chi-Square	Pr > ChiSq
sex	1	1.12	0.2901
origin	1	14.57	0.0001
type	1	0.85	0.3565
grade	3	11.94	0.0076

Display 9.4

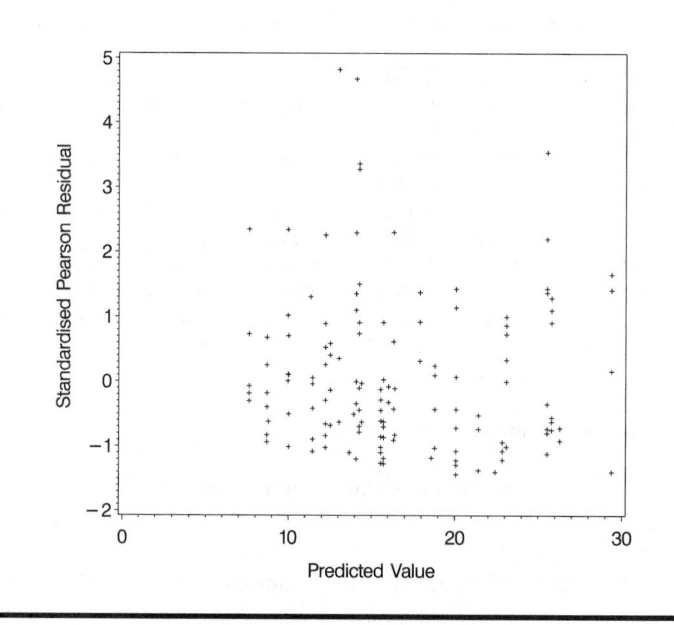

Display 9.5

Exercises

9.1 Test the significance of the interaction between class and race when using a Poisson model for the Australian school children data.

9.2 Dichotomise days absent from school by classifying 14 days or more as frequently absent. Analyse this new response variable using both the logistic and probit link and the binomial family.

Chapter 10

Longitudinal Data I: The Treatment of Postnatal Depression

10.1 Description of Data

The data set to be analysed in this chapter originates from a clinical trial of the use of oestrogen patches in the treatment of postnatal depression. Full details of the study are given in Gregoire et al. (1998). In total, 61 women with major depression, which began within 3 months of childbirth and persisted for up to 18 months postnatally, were allocated randomly to the active treatment or a placebo (a dummy patch); 34 received the former and the remaining 27 received the latter. The women were assessed twice pretreatment and then monthly for 6 months after treatment on the Edinburgh postnatal depression scale (EPDS), higher values of which indicate increasingly severe depression. The data are shown in Display 10.1. A value of –9 in this table indicates that the corresponding observation was not made for some reason.

```
0.  18  18  17  18  15  17  14  15
0.  25  27  26  23  18  17  12  10
0.  19  16  17  14  -9  -9  -9  -9
0.  24  17  14  23  17  13  12  12
0.  19  15  12  10   8   4   5   5
0.  22  20  19  11   9   8   6   5
0.  28  16  13  13   9   7   8   7
0.  24  28  26  27  -9  -9  -9  -9
0.  27  28  26  24  19  13  11   9
0.  18  25   9  12  15  12  13  20
0.  23  24  14  -9  -9  -9  -9  -9
0.  21  16  19  13  14  23  15  11
0.  23  26  13  22  -9  -9  -9  -9
0.  21  21   7  13  -9  -9  -9  -9
0.  22  21  18  -9  -9  -9  -9  -9
0.  23  22  18  -9  -9  -9  -9  -9
0.  26  26  19  13  22  12  18  13
0.  20  19  19   7   8   2   5   6
0.  20  22  20  15  20  17  15  13
0.  15  16   7   8  12  10  10  12
0.  22  21  19  18  16  13  16  15
0.  24  20  16  21  17  21  16  18
0.  -9  17  15  -9  -9  -9  -9  -9
0.  24  22  20  21  17  14  14  10
0.  24  19  16  19  -9  -9  -9  -9
0.  22  21   7   4   4   4   3   3
0.  16  18  19  -9  -9  -9  -9  -9
1.  21  21  13  12   9   9  13   6
1.  27  27   8  17  15   7   5   7
1.  24  15   8  12  10  10   6   5
1.  28  24  14  14  13  12  18  15
1.  19  15  15  16  11  14  12   8
1.  17  17   9   5   3   6   0   2
1.  21  20   7   7   7  12   9   6
1.  18  18   8   1   1   2   0   1
1.  24  28  11   7   3   2   2   2
1.  21  21   7   8   6   6   4   4
1.  19  18   8   6   4  11   7   6
1.  28  27  22  27  24  22  24  23
1.  23  19  14  12  15  12   9   6
1.  21  20  13  10   7   9  11  11
1.  18  16  17  26  -9  -9  -9  -9
1.  22  21  19   9   9  12   5   7
1.  24  23  11   7   5   8   2   3
1.  23  23  16  13  -9  -9  -9  -9
1.  24  24  16  15  11  11  11  11
1.  25  25  20  18  16   9  10   6
```

1.	15	22	15	17	12	9	8	6
1.	26	20	7	2	1	0	0	2
1.	22	20	12	8	6	3	2	3
1.	24	25	15	24	18	15	13	12
1.	22	18	17	6	2	2	0	1
1.	27	26	1	18	10	13	12	10
1.	22	20	27	13	9	8	4	5
1.	20	17	20	10	8	8	7	6
1.	22	22	12	-9	-9	-9	-9	-9
1.	20	22	15	2	4	6	3	3
1.	21	23	11	9	10	8	7	4
1.	17	17	15	-9	-9	-9	-9	-9
1.	18	22	7	12	15	-9	-9	-9
1.	23	26	24	-9	-9	-9	-9	-9

0 = Placebo, 1 = Active

Display 10. 1

10.2 The Analyses of Longitudinal Data

The data in Display 10.1 consist of repeated observations over time on each of the 61 patients; they are a particular form of repeated measures data (see Chapter 7), with time as the single within-subjects factor. The analysis of variance methods described in Chapter 7 could be, and frequently are, applied to such data; but in the case of longitudinal data, the sphericity assumption is very unlikely to be plausible — observations closer together in time are very likely more highly correlated than those taken further apart. Consequently, other methods are generally more useful for this type of data. This chapter considers a number of relatively simple approaches, including:

- Graphical displays
- Summary measure or response feature analysis

Chapter 11 discusses more formal modelling techniques that can be used to analyse longitudinal data.

10.3 Analysis Using SAS

Data sets for longitudinal and repeated measures data can be structured in two ways. In the first form, there is one observation (or case) per subject and the repeated measurements are held in separate variables. Alternatively, there may be separate observations for each measurement, with variables indicating which subject and occasion it belongs to. When analysing longitudinal data, both formats may be needed. This is typically achieved by reading the raw data into a data set in one format and then using a second data step to reformat it. In the example below, both types of data set are created in the one data step.

We assume that the data are in an ASCII file 'channi.dat' in the current directory and that the data values are separated by spaces.

```
data pndep(keep=idno group x1-x8) pndep2(keep=idno group
time dep);
   infile 'channi.dat';
   input group x1-x8;
   idno=_n_;
   array xarr {8} x1-x8;
   do i=1 to 8;
      if xarr{i}=-9 then xarr{i}=.;
         time=i;
         dep=xarr{i};
         output pndep2;
   end;
   output pndep;
run;
```

The data statement contains the names of two data sets, pndep and pndep2, indicating that two data sets are to be created. For each, the keep= option in parentheses specifies which variables are to be retained in each. The input statement reads the group information and the eight depression scores. The raw data comprise 61 such lines, so the automatic SAS variable _n_ will increment from 1 to 61 accordingly. The variable idno is assigned its value to use as a case identifier because _n_ is not stored in the data set itself.

The eight depression scores are declared an array and a do loop processes them individually. The value –9 in the data indicates a missing value and these are reassigned to the SAS missing value by the if-then statement. The variable time records the measurement occasion as 1 to 8 and dep contains the depression score for that occasion. The output statement writes an observation to the data set pndep2. From the data statement we can see that this data set will contain the subject identifier, idno, plus group, time, and dep. Because this output statement is within the do loop, it will be executed for each iteration of the do loop (i.e., eight times).

The second output statement writes an observation to the pndep data set. This data set contains idno, group, and the eight depression scores in the variables x1 to x8.

Having run this data step, the SAS log confirms that pndep has 61 observations and pndep2 488 (i.e., 61 × 8).

To begin, let us look at some means and variances of the observations. Proc means, proc summary, or proc univariate could all be used for this, but proc tabulate gives particularly neat and concise output. The second, one case per measurement, format allows a simpler specification of the table, which is shown in Display 10.2.

```
proc tabulate data=pndep2 f=6.2;
   class group time;
   var dep;
   table time,
         group*dep*(mean var n);
run;
```

	group					
	0			1		
	dep			dep		
	Mean	Var	N	Mean	Var	N
time						
1	21.92	10.15	26.00	21.94	10.54	34.00
2	20.78	15.64	27.00	21.24	12.61	34.00
3	16.48	27.87	27.00	13.35	30.84	34.00
4	15.86	37.74	22.00	11.71	43.01	31.00
5	14.12	24.99	17.00	9.10	30.02	29.00
6	12.18	34.78	17.00	8.80	21.71	28.00
7	11.35	20.24	17.00	7.29	33.10	28.00
8	10.82	22.15	17.00	6.46	22.48	28.00

Display 10.2

There is a general decline in the EPDS over time in both groups, with the values in the active treatment group (group = 1) being consistently lower.

10.3.1 Graphical Displays

Often, a useful preliminary step in the analysis of longitudinal data is to graph the observations in some way. The aim is to highlight two particular aspects of the data: how they evolve over time and how the measurements made at different times are related. A number of graphical displays might be helpful here, including:

- Separate plots of each subject's response against time, differentiating in some way between subjects in different groups
- Box plots of the observations at each time point for each group
- A plot of means and standard errors by treatment group for every time point
- A scatterplot matrix of the repeated measurements

These plots can be produced in SAS as follows.

```
symbol1 i=join v=none l=1 r=27;
symbol2 i=join v=none l=2 r=34;
proc gplot data=pndep2;
   plot dep*time=idno /nolegend skipmiss;
run;
```

Plot statements of the form plot y*x=z were introduced in Chapter 1. To produce a plot with a separate line for each subject, the subject identifier idno is used as the z variable. Because there are 61 subjects, this implies 61 symbol definitions, but it is only necessary to distinguish the two treatment groups. Thus, two symbol statements are defined, each specifying a different line type, and the repeat (r=) option is used to replicate that symbol the required number of times for the treatment group. In this instance, the data are already in the correct order for the plot. Otherwise, they would need to be sorted appropriately.

The graph (Display 10.3), although somewhat "messy," demonstrates the variability in the data, but also indicates the general decline in the depression scores in both groups, with those in the active group remaining generally lower.

```
proc sort data=pndep2;
   by group time;

proc boxplot data=pndep2;
   plot dep*time;
   by group;
run;
```

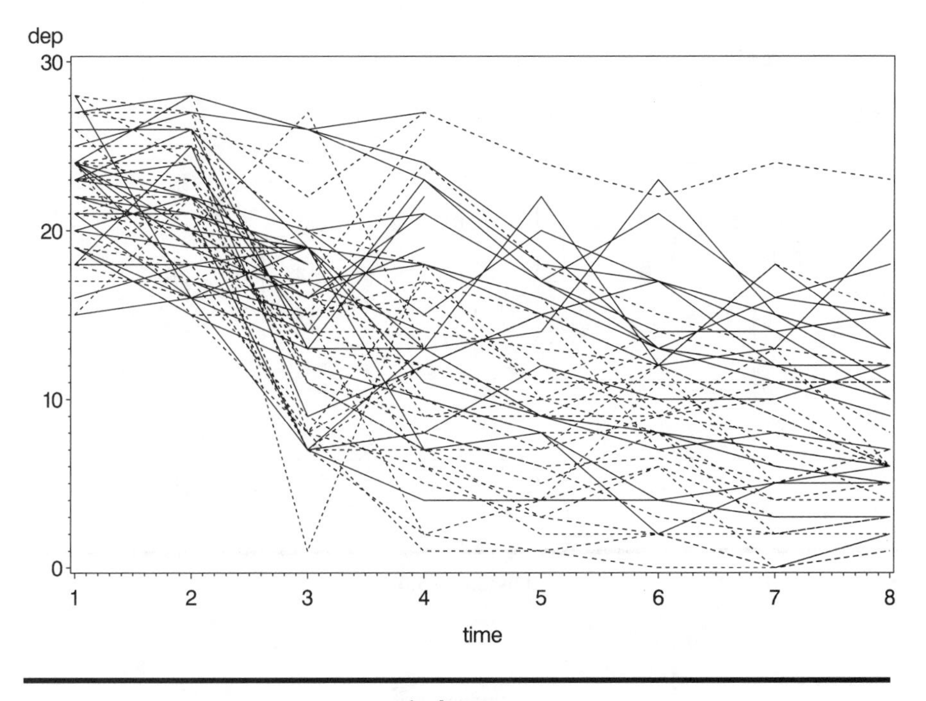

Display 10.3

The data are first sorted by **group** and **time** within group. To use the **by** statement to produce separate box plots for each group, the data must be sorted by **group**. **Proc boxplot** also requires the data to be sorted by the x-axis variable, **time** in this case. The results are shown in Displays 10.4 and 10.5. Again, the decline in depression scores in both groups is clearly seen in the graphs.

```
goptions reset=symbol;
symbol1 i=stdm1j l=1;
symbol2 i=stdm1j l=2;
proc gplot data=pndep2;
  plot dep*time=group;
run;
```

The **goptions** statement resets symbols to their defaults and is recommended when redefining **symbol** statements that have been previously used in the same session. The **std** interpolation option can be used to plot means and their standard deviations for data where multiple values of y occur for each value of x: **std1**, **std2**, and **std3** result in a line 1, 2, and 3 standard deviations above and below the mean. Where **m** is suffixed,

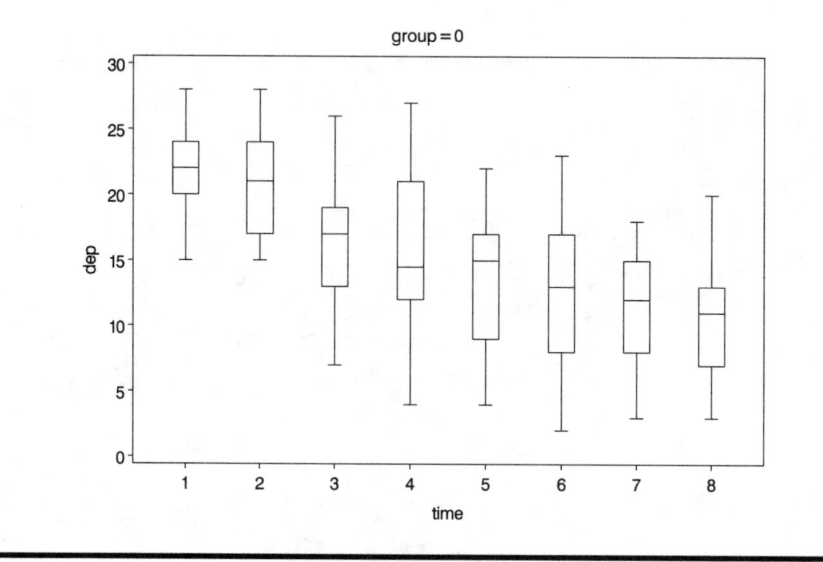

Display 10.4

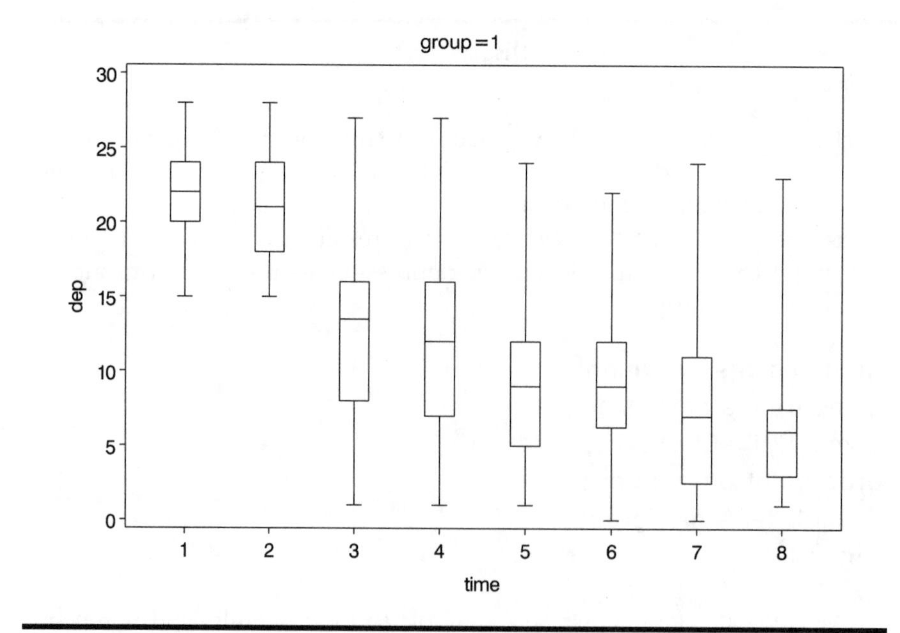

Display 10.5

as here, it is the standard error of the mean that is used. The j suffix specifies that the means should be joined. There are two groups in the data, so two **symbol** statements are used with different l (**linetype**) options to distinguish them. The result is shown in Display 10.6, which shows that from the first visit after randomisation (time 3), the depression score in the active group is lower than in the control group, a situation that continues for the remainder of the trial.

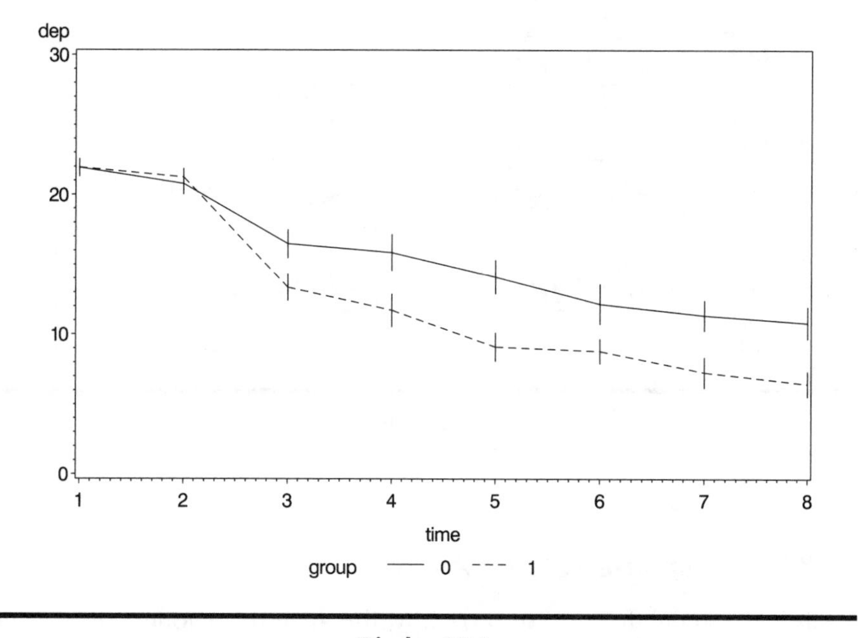

Display 10.6

The scatterplot matrix is produced using the **scattmat** SAS macro introduced in Chapter 4 and listed in Appendix A. The result is shown in Display 10.7. Clearly, observations made on occasions close together in time are more strongly related than those made further apart, a phenomenon that may have implications for more formal modelling of the data (see Chapter 11).

```
%include 'scattmat.sas';
%scattmat(pndep,x1-x8);
```

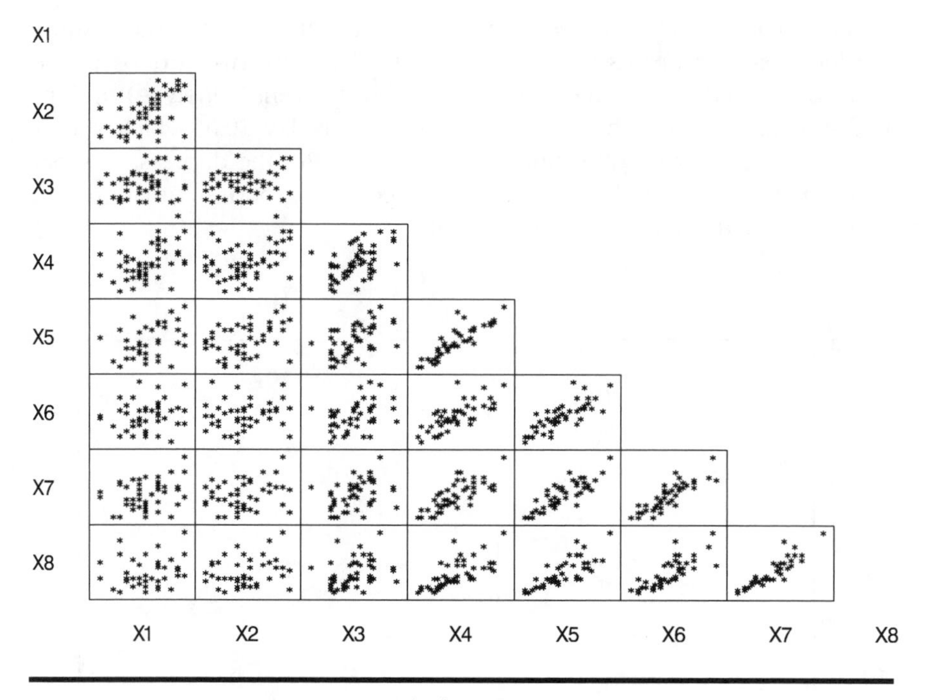

Display 10.7

10.3.2 Response Feature Analysis

A relatively straightforward approach to the analysis of longitudinal data is that involving the use of summary measures, sometimes known as *response feature analysis*. The repeated observations on a subject are used to construct a single number that characterises some relevant aspect of the subject's response profile. (In some situations, more than a single summary measure may be needed to characterise the profile adequately.) The summary measure to be used does, of course, need to be decided upon prior to the analysis of the data.

The most commonly used summary measure is the mean of the responses over time because many investigations (e.g., clinical trials) are most concerned with differences in overall level rather than more subtle effects. However, other summary measures might be considered more relevant in particular circumstances, and Display 10.8 lists a number of alternative possibilities.

Type of Data	Questions of Interest	Summary Measure
Peaked	Is overall value of outcome variable the same in different groups?	Overall mean (equal time intervals) or area under curve (unequal intervals)
Peaked	Is maximum (minimum) response different between groups?	Maximum (minimum) value
Peaked	Is time to maximum (minimum) response different between groups?	Time to maximum (minimum) response
Growth	Is rate of change of outcome different between groups?	Regression coefficient
Growth	Is eventual value of outcome different between groups?	Final value of outcome or difference between last and first values or percentage change between first and last values
Growth	Is response in one group delayed relative to the other?	Time to reach a particular value (e.g., a fixed percentage of baseline)

Display 10.8

Having identified a suitable summary measure, the analysis of the repeated measures data reduces to a simple univariate test of group differences on the chosen measure. In the case of two groups, this will involve the application of a two-sample *t*-test or perhaps its nonparametric equivalent.

Returning to the oestrogen patch data, we will use the mean as the chosen summary measure, but there are two further problems to consider:

1. How to deal with the missing values
2. How to incorporate the pretreatment measurements into an analysis

The missing values can be dealt with in at least three ways:

1. Take the mean over the available observations for a subject; that is, if a subject has only four post-treatment values recorded, use the mean of these.

2. Include in the analysis only those subjects with all six post-treatment observations.
3. Impute the missing values in some way; for example, use the last observation carried forward (LOCF) approach popular in the pharmaceutical industry.

The pretreatment values might be incorporated by calculating change scores, that is, post-treatment mean – pretreatment mean value, or as covariates in an analysis of covariance of the post-treatment means. Let us begin, however, by simply ignoring the pretreatment values and deal only with the post-treatment means.

The three possibilities for calculating the mean summary measure can be implemented as follows:

```
data pndep;
   set pndep;
   array xarr {8} x1-x8;
   array locf  {8} locf1-locf8;
   do i=3 to 8;
      locf{i}=xarr{i};
      if xarr{i}=. then locf{i}=locf{i-1};
   end;
   mnbase=mean(x1,x2);
   mnresp=mean(of x3-x8);
   mncomp=(x3+x4+x5+x6+x7+x8)/6;
   mnlocf=mean(of locf3-locf8);
   chscore=mnbase-mnresp;
run;
```

The summary measures are to be included in the **pndep** data set, so this is named in the **data** statement. The **set** statement indicates that the data are to be read from the current version of **pndep**. The eight depression scores **x1-x8** are declared as an array and another array is declared for the LOCF values. Eight variables are declared, although only six will be used. The **do** loop assigns LOCF values for those occasions when the depression score was missing. The mean of the two baseline measures is then computed using the SAS **mean** function. The next statement computes the mean of the recorded follow-up scores. When a variable list is used with the mean function, it must be preceded with **'of'**. The **mean** function will only result in a missing value if *all* the variables are missing. Otherwise, it computes the mean of the non-missing values. Thus, the **mnresp** variable will contain the mean of the available follow-up scores for a subject. Because an arithmetic operation involving a missing value results in a missing value, **mncomp** will be assigned a missing value if *any* of the variables is missing.

A *t*-test can now be applied to assess difference between treatments for each of the three procedures. The results are shown in Display 10.9.

```
proc ttest data=pndep;
   class group;
   var mnresp mnlocf mncomp;
run;
```

The TTEST Procedure

Statistics

Variable	group	N	Lower CL Mean	Mean	Upper CL Mean	Lower CL Std Dev	Std Dev	Upper CL Std Dev	Std Err
mnresp	0	27	12.913	14.728	16.544	3.6138	4.5889	6.2888	0.8831
mnresp	1	34	8.6447	10.517	12.39	4.3284	5.3664	7.0637	0.9203
mnresp	Diff (1-2)		1.6123	4.2112	6.8102	4.2709	5.0386	6.1454	1.2988
mnlocf	0	27	13.072	14.926	16.78	3.691	4.6868	6.423	0.902
mnlocf	1	34	8.6861	10.63	12.574	4.4935	5.5711	7.3331	0.9554
mnlocf	Diff (1-2)		1.6138	4.296	6.9782	4.4077	5.2	6.3422	1.3404
mncomp	0	17	11.117	13.333	15.55	3.2103	4.3104	6.5602	1.0454
mncomp	1	28	7.4854	9.2589	11.032	3.616	4.5737	6.2254	0.8643
mncomp	Diff (1-2)		1.298	4.0744	6.8508	3.6994	4.4775	5.6731	1.3767

T-Tests

Variable	Method	Variances	DF	t Value	Pr > |t|
mnresp	Pooled	Equal	59	3.24	0.0020
mnresp	Satterthwaite	Unequal	58.6	3.30	0.0016
mnlocf	Pooled	Equal	59	3.20	0.0022
mnlocf	Satterthwaite	Unequal	58.8	3.27	0.0018
mncomp	Pooled	Equal	43	2.96	0.0050
mncomp	Satterthwaite	Unequal	35.5	3.00	0.0049

Equality of Variances

Variable	Method	Num DF	Den DF	F Value	Pr > F
mnresp	Folded F	33	26	1.37	0.4148
mnlocf	Folded F	33	26	1.41	0.3676
mncomp	Folded F	27	16	1.13	0.8237

Display 10.9

Here, the results are similar and the conclusion in each case the same; namely, that there is a substantial difference in overall level in the two treatment groups. The confidence intervals for the treatment effect given by each of the three procedures are:

- Using mean of available observations (1.612, 6.810)
- Using LOCF (1.614, 6.987)
- Using only complete cases (1.298, 6.851)

All three approaches lead, in this example, to the conclusion that the active treatment considerably lowers depression. But, in general, using only subjects with a complete set of measurements and last observation carried forward are not to be recommended. Using only complete observations can produce bias in the results unless the missing observations are *missing completely at random* (see Everitt and Pickles [1999]). And the LOCF procedure has little in its favour because it makes highly unlikely assumptions; for example, that the expected value of the (unobserved) remaining observations remain at their last recorded value. Even using the mean of the values actually recorded is not without its problems (see Matthews [1993]), but it does appear, in general, to be the least objectionable of the three alternatives.

Now consider analyses that make use of the pretreatment values available for each woman in the study. The change score analysis and the analysis of covariance using the mean of available post-treatment values as the summary and the mean of the two pretreatment values as covariate can be applied as follows:

```
proc glm data=pndep;
   class group;
   model chscore=group /solution;

proc glm data=pndep;
   class group;
   model mnresp=mnbase group /solution;
run;
```

We use proc glm for both analyses for comparability, although we could also have used a t-test for the change scores. The results are shown in Display 10.10. In both cases for this example, the group effect is highly significant, confirming the difference in depression scores of the active and control group found in the previous analysis.

In general, the analysis of covariance approach is to be preferred for reasons outlined in Senn (1998) and Everitt and Pickles (2000).

```
                     The GLM Procedure

                  Class Level Information

                  Class     Levels    Values

                  group          2       0 1

               Number of observations    61

                     The GLM Procedure

Dependent Variable: chscore

                              Sum of
 Source              DF      Squares   Mean Square   F Value    Pr > F

 Model                1   310.216960    310.216960     12.17    0.0009

 Error               59  1503.337229     25.480292

 Corrected Total     60  1813.554189

        R-Square    Coeff Var    Root MSE     chscore Mean

        0.171055     55.70617    5.047801         9.061475

     Source    DF      Type I SS   Mean Square   F Value    Pr > F

     group      1    310.2169603   310.2169603     12.17    0.0009

     Source    DF    Type III SS   Mean Square   F Value    Pr > F

     group      1    310.2169603   310.2169603     12.17    0.0009

                                         Standard
      Parameter           Estimate          Error    t Value    Pr > |t|

      Intercept       11.07107843 B    0.86569068      12.79      <.0001
      group 0         -4.54021423 B    1.30120516      -3.49      0.0009
      group 1          0.00000000 B             .          .           .

NOTE: The X'X matrix has been found to be singular, and a generalized inverse
      was used to solve the normal equations.  Terms whose estimates are
      followed by the letter 'B' are not uniquely estimable.
```

```
                          The GLM Procedure

                       Class Level Information

                       Class    Levels    Values

                       group         2      0 1

                     Number of observations   61

                          The GLM Procedure

Dependent Variable: mnresp

                                    Sum of
 Source                  DF        Squares    Mean Square    F Value    Pr > F

 Model                    2     404.398082     202.199041       8.62    0.0005

 Error                   58    1360.358293      23.454453

 Corrected Total         60    1764.756375

             R-Square     Coeff Var     Root MSE    mnresp Mean

             0.229152      39.11576     4.842980       12.38115

    Source       DF       Type I SS    Mean Square    F Value    Pr > F

    mnbase        1     117.2966184    117.2966184       5.00    0.0292
    group         1     287.1014634    287.1014634      12.24    0.0009

    Source       DF     Type III SS    Mean Square    F Value    Pr > F

    mnbase        1     137.5079844    137.5079844       5.86    0.0186
    group         1     287.1014634    287.1014634      12.24    0.0009
```

Parameter	Estimate	Standard Error	t Value	Pr > |t|
Intercept	-0.171680099 B	4.49192993	-0.04	0.9696
mnbase	0.495123238	0.20448526	2.42	0.0186
group 0	4.374121879 B	1.25021825	3.50	0.0009
group 1	0.000000000 B		.	.

NOTE: The X'X matrix has been found to be singular, and a generalized inverse was used to solve the normal equations. Terms whose estimates are followed by the letter 'B' are not uniquely estimable.

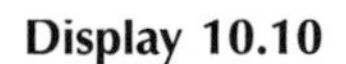

Display 10.10

Exercises

10.1 The graph in Display 10.3 indicates the phenomenon known as "tracking," the tendency of women with higher depression scores at the beginning of the trial to be those with the higher scores at the end. This phenomenon becomes more visible if standardized scores are plotted [i.e., (depression scores – visit mean)/visit S.D.]. Calculate and plot these scores, differentiating on the plot the women in the two treatment groups.

10.2 Apply the response feature approach described in the text, but now using the slope of each woman's depression score on time as the summary measure.

Chapter 11

Longitudinal Data II: The Treatment of Alzheimer's Disease

11.1 Description of Data

The data used in this chapter are shown in Display 11.1. They arise from an investigation of the use of lecithin, a precursor of choline, in the treatment of Alzheimer's disease. Traditionally, it has been assumed that this condition involves an inevitable and progressive deterioration in all aspects of intellect, self-care, and personality. Recent work suggests that the disease involves pathological changes in the central cholinergic system, which might be possible to remedy by long-term dietary enrichment with lecithin. In particular, the treatment might slow down or perhaps even halt the memory impairment associated with the condition. Patients suffering from Alzheimer's disease were randomly allocated to receive either lecithin or placebo for a 6-month period. A cognitive test score giving the number of words recalled from a previously given standard list was recorded monthly for 5 months.

The main question of interest here is whether the lecithin treatment has had any effect.

	Visit				
Group	*1*	*2*	*3*	*4*	*5*
1	20	15	14	13	13
1	14	12	12	10	10
1	7	5	5	6	5
1	6	10	9	8	7
1	9	7	9	5	4
1	9	9	9	11	8
1	7	3	7	6	5
1	18	17	16	14	12
1	6	9	9	9	9
1	10	15	12	12	11
1	5	9	7	3	5
1	11	11	8	8	9
1	10	2	9	3	5
1	17	12	14	10	9
1	16	15	12	7	9
1	7	10	4	7	5
1	5	0	5	0	0
1	16	7	7	6	4
1	2	1	1	2	2
1	7	11	7	5	8
1	9	16	14	10	6
1	2	5	6	7	6
1	7	3	5	5	5
1	19	13	14	12	10
1	7	5	8	8	6
2	9	12	16	17	18
2	6	7	10	15	16
2	13	18	14	21	21
2	9	10	12	14	15
2	6	7	8	9	12
2	11	11	12	14	16
2	7	10	11	12	14
2	8	18	19	19	22
2	3	3	3	7	8
2	4	10	11	17	18
2	11	10	10	15	16
2	1	3	2	4	5
2	6	7	7	9	10
2	0	3	3	4	6

2	18	18	19	22	22
2	15	15	15	18	19
2	10	14	16	17	19
2	6	6	7	9	10
2	9	9	13	16	20
2	4	3	4	7	9
2	4	13	13	16	19
2	10	11	13	17	21

1 = Placebo, 2 = Lecithin

Display 11.1

11.2 Random Effects Models

Chapter 10 considered some suitable graphical methods for longitudinal data, and a relatively straightforward inferential procedure. This chapter considers a more formal modelling approach that involves the use of *random effects models*. In particular, we consider two such models: one that allows the participants to have different intercepts of cognitive score on time, and the other that also allows the possibility of the participants having different slopes for the regression of cognitive score on time.

Assuming that y_{ijk} represents the cognitive score for subject k on visit j in group i, the random intercepts model is

$$y_{ijk} = (\beta_0 + a_k) + \beta_1 \text{ Visit}_j + \beta_2 \text{ Group}_i + \in_{ijk} \tag{11.1}$$

where β_0, β_1, and β_2 are respectively the intercept and regression coefficients for Visit and Group (where Visit takes the values 1, 2, 3, 4, and 5, and Group the values 1 for placebo and 2 for lecithin); a_k are random effects that model the shift in intercept for each subject, which because there is a fixed change for visit, are preserved for all values of visit; and the $\in_{ijk}$ are residual or error terms. The a_k are assumed to have a normal distribution with mean zero and variance σ_a^2. The $\in_{ijk}$ are assumed to have a normal distribution with mean zero and variance σ^2. Such a model implies a compound symmetry covariance pattern for the five repeated measures (see Everitt [2001] for details).

The model allowing for both random intercept and random slope can be written as:

$$y_{ijk} = (\beta_0 + a_k) + (\beta_1 + b_k) \text{ Visit}_j + \beta_2 \text{ Group}_i + \in_{ijk} \tag{11.2}$$

Now a further random effect has been added to the model compared to Eq. (11.1). The terms b_k are assumed to be normally distributed with mean zero and variance σ_b^2. In addition, the possibility that the random effects are not independent is allowed for by introducing a covariance term for them, σ_{ab}.

The model in Eq. (11.2) can be conveniently written in matrix notation as:

$$\mathbf{y}_{ik} = \mathbf{X}_i\boldsymbol{\beta} + \mathbf{Z}\mathbf{b}_k + \boldsymbol{\epsilon}_{ik} \tag{11.3}$$

where now

$$\mathbf{Z} = \begin{bmatrix} 1 & 1 \\ 1 & 2 \\ 1 & 3 \\ 1 & 4 \\ 1 & 5 \end{bmatrix}$$

$$\mathbf{b}_k' = [a_k, b_k]$$

$$\boldsymbol{\beta}' = [\beta_0, \beta_1, \beta_2]$$

$$\mathbf{y}_{ik}' = [y_{i1k}, y_{i2k}, y_{i3k}, y_{i4k}, y_{i5k}]$$

$$\mathbf{X}_i = \begin{bmatrix} 1 & 1 & \text{Group}_i \\ 1 & 2 & \text{Group}_i \\ 1 & 3 & \text{Group}_i \\ 1 & 4 & \text{Group}_i \\ 1 & 5 & \text{Group}_i \end{bmatrix}$$

$$\boldsymbol{\epsilon}_{ik}' = [\epsilon_{i1k}, \epsilon_{i2k}, \epsilon_{i3k}, \epsilon_{i4k}, \epsilon_{i5k}]$$

The model implies the following covariance matrix for the repeated measures:

$$\boldsymbol{\Sigma} = \mathbf{Z}\boldsymbol{\Psi}\mathbf{Z}' + \sigma^2\mathbf{I} \tag{11.4}$$

where

$$\Psi = \begin{bmatrix} \sigma_a^2 & \sigma_{ab} \\ \sigma_{ab} & \sigma_b^2 \end{bmatrix}$$

Details of how to fit such models are given in Pinheiro and Bates (2000).

11.3 Analysis Using SAS

We assume that the data shown in Display 11.1 are in an ASCII file, **alzheim.dat**, in the current directory. The data step below reads the data and creates a SAS data set **alzheim**, with one case per measurement. The grouping variable and five monthly scores for each subject are read in together, and then split into separate o bservations using the **array**, iterative **do** loop, and **output** statement. This technique is described in more detail in Chapter 5. The automatic SAS variable **_n_** is used to form a subject identifier. With 47 subjects and 5 visits each, the resulting data set contains 235 observations.

```
data alzheim;
   infile 'alzheim.dat';
   input group score1-score5;
   array sc {5} score1-score5;
   idno=_n_;
   do visit=1 to 5;
      score=sc{visit};
      output;
   end;
run;
```

We begin with some plots of the data. First, the data are sorted by **group** so that the **by** statement can be used to produce separate plots for each group.

```
proc sort data=alzheim;
   by group;
run;

symbol1 i=join v=none r=25;
proc gplot data=alzheim;
   plot score*visit=idno / nolegend;
```

```
   by group;
run;
```

To plot the scores in the form of a line for each subject, we use **plot score*visit=idno**. There are 25 subjects in the first group and 22 in the second. The plots will be produced separately by group, so the symbol definition needs to be repeated 25 times and the **r=25** options does this. The plots are shown in Displays 11.2 and 11.3.

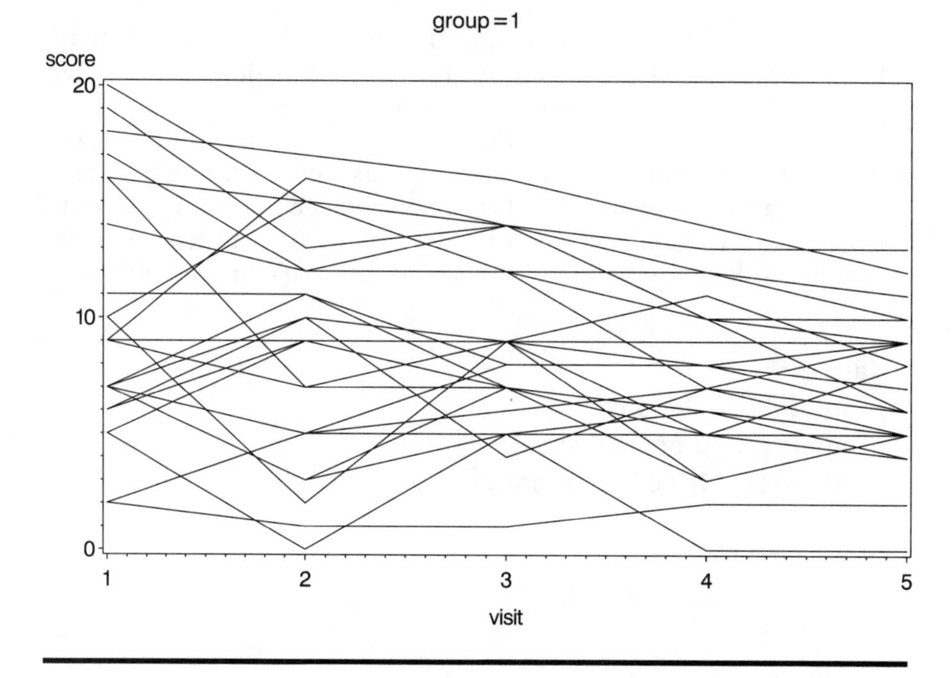

Display 11.2

Next we plot mean scores with their standard errors for each group on the same plot. (See Chapter 10 for an explanation of the following SAS statements.) The plot is shown in Display 11.4.

```
goptions reset=symbol;
symbol1 i=std1mj v=none l=1;
symbol2 i=std1mj v=none l=3;
proc gplot data=alzheim;
   plot score*visit=group;
run;
```

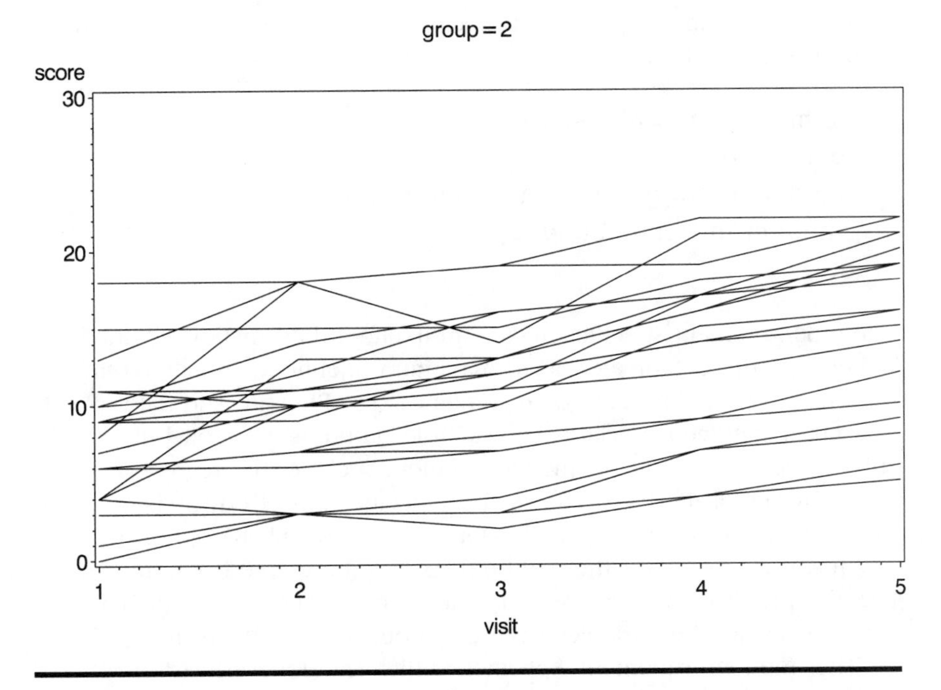

Display 11.3

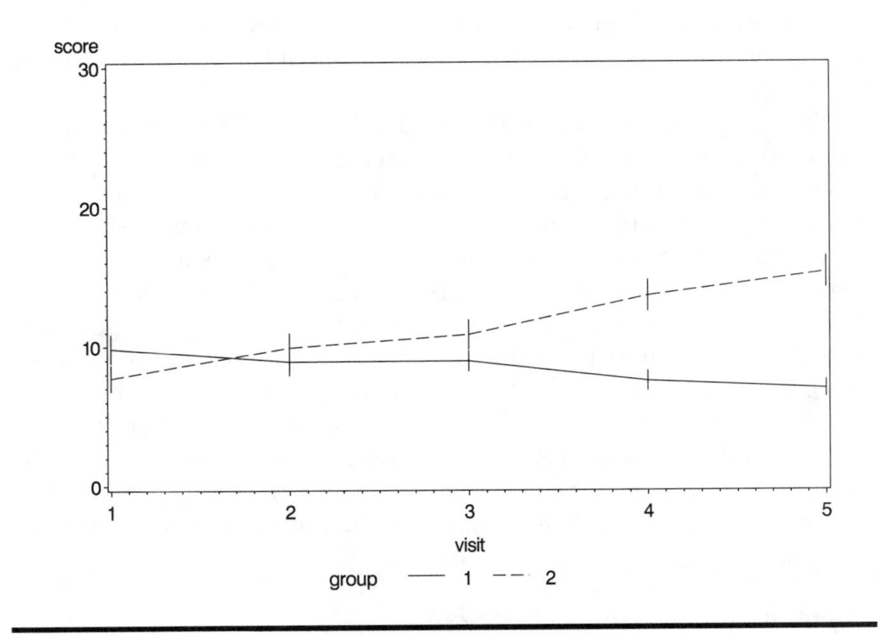

Display 11.4

The random intercepts model specified in Eq. (11.1) can be fitted using proc mixed, as follows:

```
proc mixed data=alzheim method=ml;
   class group idno;
   model score=group visit /s outpred=mixout;
   random int /subject=idno;
run;
```

The proc statement specifies maximum likelihood estimation (method=ml) rather than the default, restricted maximum likelihood (method=reml), as this enables nested models to be compared (see Pinheiro and Bates, 2000). The class statement declares the variable group as a factor, but also the subject identifier idno. The model statement specifies the regression equation in terms of the fixed effects. The specification of effects is the same as for proc glm described in Chapter 6. The s (solution) option requests parameter estimates for the fixed effects and the outpred option specifies that the predicted values are to be saved in a data set mixout. This will also contain all the variables from the input data set alzheim.

The random statement specifies which random effects are to be included in the model. For the random intercepts model, int (or intercept) is specified. The subject= option names the variable that identifies the subjects in the data set. If the subject identifier, idno in this case, is not declared in the class statement, the data set should be sorted into subject identifier order.

The results are shown in Display 11.5. We see that the parameters σ_a^2 and σ^2 are estimated to be 15.1284 and 8.2462, respectively (see "Covariance Parameter Estimates"). The tests for the fixed effects in the model indicate that both group and visit are significant. The parameter estimate for group indicates that group 1 (the placebo group) has a lower average cognitive score. The estimated treatment effect is –3.06, with a 95% confidence interval of $-3.06 \pm 1.96 \times 1.197$, that is, (–5.41, –0.71). The goodness-of-fit statistics given in Display 11.5 can be used to compare models (see later). In particular the AIC (Akaike's Information Criterion) tries to take into account both the statistical goodness-of-fit and the number of parameters needed to achieve this fit by imposing a penalty for increasing the number of parameters (for more details, see Krzanowski and Marriott [1995]).

Line plots of the predicted values for each group can be obtained as follows:

```
symbol1 i=join v=none l=1 r=30;
proc gplot data=mixout;
   plot pred*visit=idno / nolegend;
```

```
  by group;
run;
```

The plots are shown in Displays 11.6 and 11.7.

```
                          The Mixed Procedure

                           Model Information

        Data Set                                   WORK.ALZHEIM
        Dependent Variable                                score
        Covariance Structure               Variance Components
        Subject Effect                                     idno
        Estimation Method                                    ML
        Residual Variance Method                        Profile
        Fixed Effects SE Method                     Model-Based
        Degrees of Freedom Method                   Containment

                        Class Level Information

          Class    Levels    Values

          group         2    1 2
          idno         47    1 2 3 4 5 6 7 8 9 10 11 12 13
                             14 15 16 17 18 19 20 21 22 23
                             24 25 26 27 28 29 30 31 32 33
                             34 35 36 37 38 39 40 41 42 43
                             44 45 46 47

                              Dimensions

                  Covariance Parameters            2
                  Columns in X                     4
                  Columns in Z Per Subject         1
                  Subjects                        47
                  Max Obs Per Subject              5
                  Observations Used              235
                  Observations Not Used            0
                  Total Observations             235

                           Iteration History

       Iteration    Evaluations       -2 Log Like       Criterion

               0              1    1407.53869769
               1              1    1271.71980926      0.00000000
```

```
                    Convergence criteria met.

                 Covariance Parameter Estimates

                 Cov Parm     Subject     Estimate

                 Intercept    idno         15.1284
                 Residual                   8.2462

                         Fit Statistics

                -2 Log Likelihood              1271.7
                AIC (smaller is better)        1281.7
                AICC (smaller is better)       1282.0
                BIC (smaller is better)        1291.0

                      The Mixed Procedure

                  Solution for Fixed Effects

                                     Standard
Effect       group    Estimate          Error     DF    t Value    Pr > |t|

Intercept               9.9828         0.9594     45      10.40      <.0001
group        1         -3.0556         1.1974    187      -2.55      0.0115
group        2               0              .      .          .           .
visit                   0.4936         0.1325    187       3.73      0.0003

                Type 3 Tests of Fixed Effects

                         Num     Den
             Effect       DF      DF    F Value    Pr > F

             group         1     187       6.51    0.0115
             visit         1     187      13.89    0.0003
```

Display 11.5

The predicted values for both groups under the random intercept model indicate a rise in cognitive score with time, contrary to the pattern in the observed scores (see Displays 11.2 and 11.3), in which there appears to be a decline in cognitive score in the placebo group and a rise in the lecithin group.

We can now see if the random intercepts and slopes model specified in Eq. (11.3) improve the situation. The model can be fitted as follows:

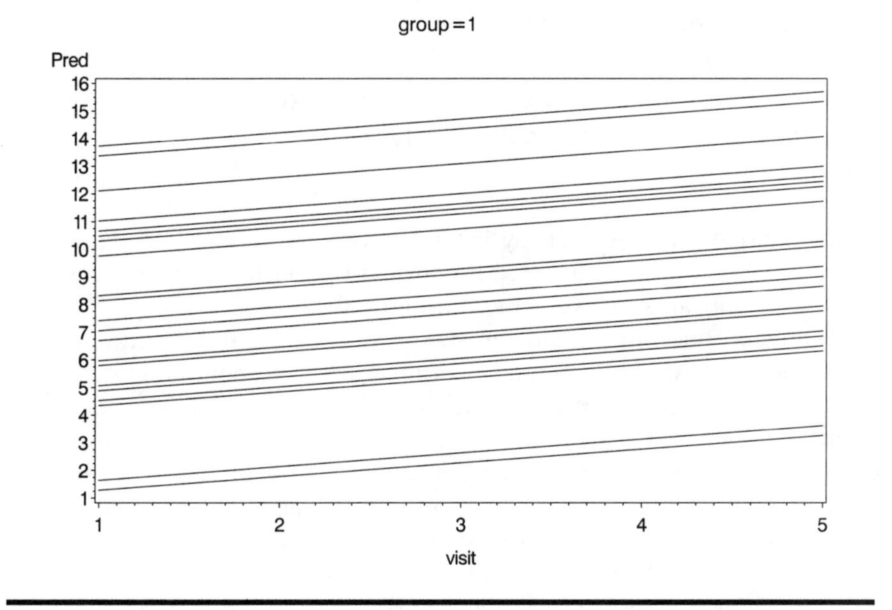

Display 11.6

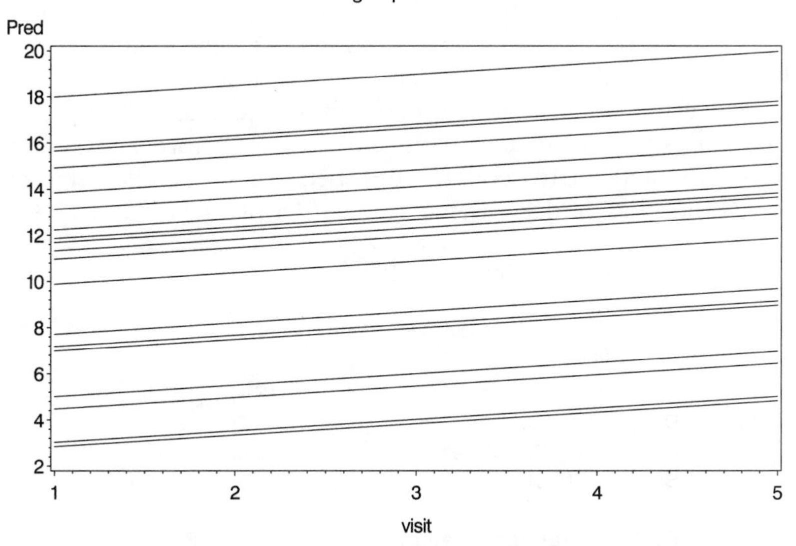

Display 11.7

```
proc mixed data=alzheim method=ml covtest;
   class group idno;
   model score=group visit /s outpred=mixout;
   random int visit /subject=idno type=un;
run;
```

Random slopes are specified by including visit on the random statement. There are two further changes. The **covtest** option in the **proc** statement requests significance tests for the random effects.

The **type** option in the **random** statement specifies the structure of the covariance matrix of the parameter estimates for the random effects. The default structure is **type=vc** (variance components), which models a different variance component for each random effect, but constrains the covariances to zero. Unstructured covariances, **type=un**, allow a separate estimation of each element of the covariance matrix. In this example, it allows an intercept-slope covariance to be estimated as a random effect, whereas the default would constrain this to be zero.

The results are shown in Display 11.8. First, we see that σ_a^2, σ_b^2, σ_{ab}, and σ^2 are estimated to be 38.7228, 2.0570, –6.8253, and 3.1036, respectively. All are significantly different from zero. The estimated correlation between intercepts and slopes resulting from these values is –0.76. Again, both fixed effects are found to be significant. The estimated treatment effect, –3.77, is very similar to the value obtained with the random-intercepts-only model. Comparing the AIC values for the random intercepts model (1281.7) and the random intercepts and slopes model (1197.4) indicates that the latter provides a better fit for these data. The predicted values for this second model, plotted exactly as before and shown in Displays 11.9 and 11.10, confirm this because they reflect far more accurately the plots of the observed data in Displays 11.2 and 11.3.

```
                    The Mixed Procedure

                     Model Information

Data Set                              WORK.ALZHEIM
Dependent Variable                           score
Covariance Structure                  Unstructured
Subject Effect                                idno
Estimation Method                               ML
Residual Variance Method                   Profile
Fixed Effects SE Method                Model-Based
Degrees of Freedom Method              Containment
```

```
                Class Level Information

      Class    Levels    Values

      group        2     1 2
      idno        47     1 2 3 4 5 6 7 8 9 10 11 12 13
                         14 15 16 17 18 19 20 21 22 23
                         24 25 26 27 28 29 30 31 32 33
                         34 35 36 37 38 39 40 41 42 43
                         44 45 46 47

                        Dimensions

           Covariance Parameters            4
           Columns in X                     4
           Columns in Z Per Subject         2
           Subjects                        47
           Max Obs Per Subject              5
           Observations Used              235
           Observations Not Used            0
           Total Observations             235

                     Iteration History

     Iteration    Evaluations      -2 Log Like       Criterion

             0              1    1407.53869769
             1              2    1183.35682556    0.00000349
             2              1    1183.35550402    0.00000000

                 Convergence criteria met.

              Covariance Parameter Estimates

                                           Standard       Z
  Cov Parm     Subject     Estimate        Error      Value      Pr Z

  UN(1,1)      idno         38.7228      12.8434       3.01    0.0013
  UN(2,1)      idno         -6.8253       2.3437      -2.91    0.0036
  UN(2,2)      idno          2.0570       0.4898       4.20    <.0001
  Residual                   3.1036       0.3696       8.40    <.0001
```

Fit Statistics

-2 Log Likelihood	1183.4
AIC (smaller is better)	1197.4
AICC (smaller is better)	1197.8
BIC (smaller is better)	1210.3

The Mixed Procedure

Null Model Likelihood Ratio Test

DF	Chi-Square	Pr > ChiSq
3	224.18	<.0001

Solution for Fixed Effects

Effect	group	Estimate	Standard Error	DF	t Value	Pr > |t|
Intercept		10.3652	1.1406	45	9.09	<.0001
group	1	-3.7745	1.1955	141	-3.16	0.0019
group	2	0	.	.	.	.
visit		0.4936	0.2244	46	2.20	0.0329

Type 3 Tests of Fixed Effects

Effect	Num DF	Den DF	F Value	Pr > F
group	1	141	9.97	0.0019
visit	1	46	4.84	0.0329

Display 11.8

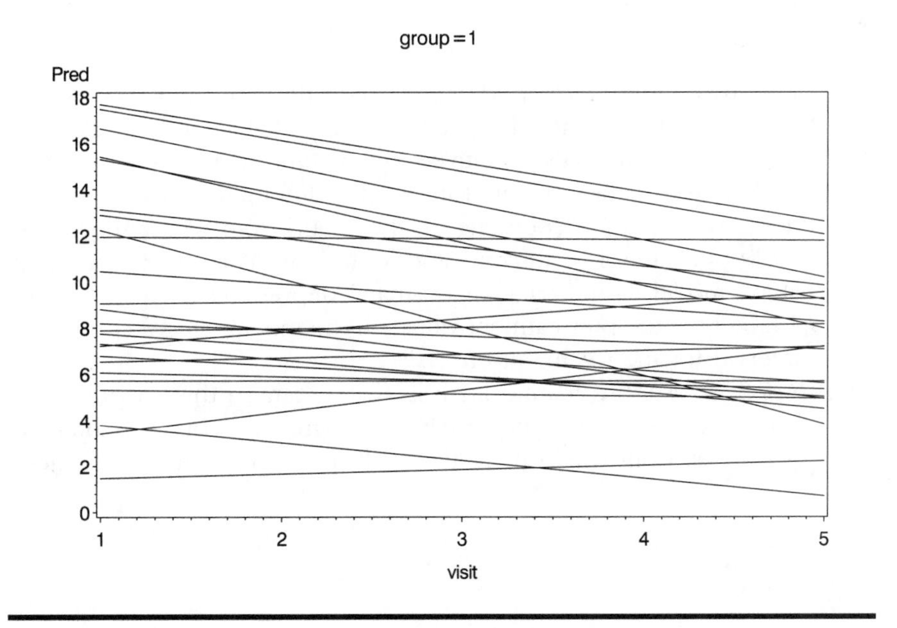

Display 11.9

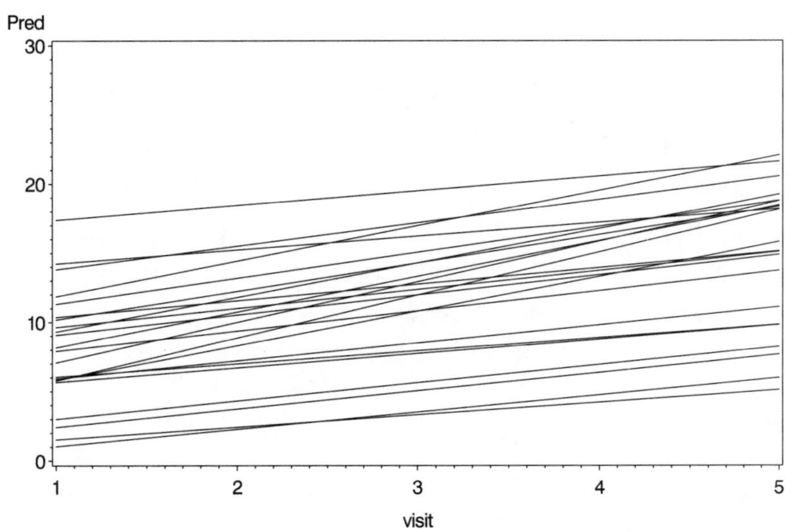

Display 11.10

Exercises

11.1 Investigate the effect of adding a fixed effect for the Group × Visit interaction to the models specified in Eqs. (11.1) and (11.2).

11.2 Regress each subject's cognitive score on time and plot the estimated slopes against the estimated intercepts, differentiating the observations by *P* and *L*, depending on the group from which they arise.

11.3 Fit both a random intercepts and a random intercepts and random slope model to the data on postnatal depression used in Chapter 10. Include the pretreatment values in the model. Find a confidence interval for the treatment effect.

11.4 Apply a response feature analysis to the data in this chapter using both the mean and the maximum cognitive score as summary measures. Compare your results with those given in this chapter.

Chapter 12

Survival Analysis: Gastric Cancer and Methadone Treatment of Heroin Addicts

12.1 Description of Data

In this chapter we analyse two data sets. The first, shown in Display 12.1, involves the survival times of two groups of 45 patients suffering from gastric cancer. Group 1 received chemotherapy and radiation, group 2 only chemotherapy. An asterisk denotes *censoring*, that is, the patient was still alive at the time the study ended. Interest lies in comparing the survival times of the two groups. (These data are given in Table 467 of *SDS*.)

However, "survival times" do not always involve the endpoint death. This is so for the second data set considered in this chapter and shown in Display 12.2. Given in this display are the times that heroin addicts remained in a clinic for methadone maintenance treatment. Here, the endpoint of interest is not death, but termination of treatment. Some subjects were still in the clinic at the time these data were recorded and this is indicated by the variable status, which is equal to 1 if the person had departed the clinic on completion of treatment and 0 otherwise.

Possible explanatory variables for time to complete treatment are maximum methadone dose, whether the addict had a criminal record, and the clinic in which the addict was being treated. (These data are given in Table 354 of *SDS*.)

For the gastric cancer data, the primary question of interest is whether or not the survival time differs in the two treatment groups; and for the methadone data, the possible effects of the explanatory variables on time to completion of treatment are of concern. It might be thought that such questions could be addressed by some of the techniques covered in previous chapters (e.g., *t*-tests or multiple regression). Survival times, however, require special methods of analysis for two reasons:

1. They are restricted to being positive so that familiar parametric assumptions (e.g., normality) may not be justifiable.
2. The data often contain censored observations, that is, observations for which, at the end of the study, the event of interest (death in the first data set, completion of treatment in the second) has not occurred; all that can be said about a censored survival time is that the unobserved, uncensored value would have been greater than the value recorded.

Group 1			*Group 2*		
17	185	542	1	383	778
42	193	567	63	383	786
44	195	577	105	388	797
48	197	580	125	394	955
60	208	795	182	408	968
72	234	855	216	460	977
74	235	1174*	250	489	1245
95	254	1214	262	523	1271
103	307	1232*	301	524	1420
108	315	1366	301	535	1460*
122	401	1455*	342	562	1516*
144	445	1585*	354	569	1551
167	464	1622*	356	675	1690*
170	484	1626*	358	676	1694
183	528	1736*	380	748	

Display 12.1

ID	*Clinic*	*Status*	*Time*	*Prison*	*Dose*	*ID*	*Clinic*	*Status*	*Time*	*Prison*	*Dose*
1	1	1	428	0	50	132	2	0	633	0	70
2	1	1	275	1	55	133	2	1	661	0	40
3	1	1	262	0	55	134	2	1	232	1	70
4	1	1	183	0	30	135	2	1	13	1	60
5	1	1	259	1	65	137	2	0	563	0	70
6	1	1	714	0	55	138	2	0	969	0	80
7	1	1	438	1	65	143	2	0	1052	0	80
8	1	0	796	1	60	144	2	0	944	1	80
9	1	1	892	0	50	145	2	0	881	0	80
10	1	1	393	1	65	146	2	1	190	1	50
11	1	0	161	1	80	148	2	1	79	0	40
12	1	1	836	1	60	149	2	0	884	1	50
13	1	1	523	0	55	150	2	1	170	0	40
14	1	1	612	0	70	153	2	1	286	0	45
15	1	1	212	1	60	156	2	0	358	0	60
16	1	1	399	1	60	158	2	0	326	1	60
17	1	1	771	1	75	159	2	0	769	1	40
18	1	1	514	1	80	160	2	1	161	0	40
19	1	1	512	0	80	161	2	0	564	1	80
21	1	1	624	1	80	162	2	1	268	1	70
22	1	1	209	1	60	163	2	0	611	1	40
23	1	1	341	1	60	164	2	1	322	0	55
24	1	1	299	0	55	165	2	0	1076	1	80
25	1	0	826	0	80	166	2	0	2	1	40
26	1	1	262	1	65	168	2	0	788	0	70
27	1	0	566	1	45	169	2	0	575	0	80
28	1	1	368	1	55	170	2	1	109	1	70
30	1	1	302	1	50	171	2	0	730	1	80
31	1	0	602	0	60	172	2	0	790	0	90
32	1	1	652	0	80	173	2	0	456	1	70
33	1	1	293	0	65	175	2	1	231	1	60
34	1	0	564	0	60	176	2	1	143	1	70
36	1	1	394	1	55	177	2	0	86	1	40
37	1	1	755	1	65	178	2	0	1021	0	80
38	1	1	591	0	55	179	2	0	684	1	80
39	1	0	787	0	80	180	2	1	878	1	60
40	1	1	739	0	60	181	2	1	216	0	100
41	1	1	550	1	60	182	2	0	808	0	60
42	1	1	837	0	60	183	2	1	268	1	40
43	1	1	612	0	65	184	2	0	222	0	40
44	1	0	581	0	70	186	2	0	683	0	100
45	1	1	523	0	60	187	2	0	496	0	40

ID	Clinic	Status	Time	Prison	Dose	ID	Clinic	Status	Time	Prison	Dose
46	1	1	504	1	60	188	2	1	389	0	55
48	1	1	785	1	80	189	1	1	126	1	75
49	1	1	774	1	65	190	1	1	17	1	40
50	1	1	560	0	65	192	1	1	350	0	60
51	1	1	160	0	35	193	2	0	531	1	65
52	1	1	482	0	30	194	1	0	317	1	50
53	1	1	518	0	65	195	1	0	461	1	75
54	1	1	683	0	50	196	1	1	37	0	60
55	1	1	147	0	65	197	1	1	167	1	55
57	1	1	563	1	70	198	1	1	358	0	45
58	1	1	646	1	60	199	1	1	49	0	60
59	1	1	899	0	60	200	1	1	457	1	40
60	1	1	857	0	60	201	1	1	127	0	20
61	1	1	180	1	70	202	1	1	7	1	40
62	1	1	452	0	60	203	1	1	29	1	60
63	1	1	760	0	60	204	1	1	62	0	40
64	1	1	496	0	65	205	1	0	150	1	60
65	1	1	258	1	40	206	1	1	223	1	40
66	1	1	181	1	60	207	1	0	129	1	40
67	1	1	386	0	60	208	1	0	204	1	65
68	1	0	439	0	80	209	1	1	129	1	50
69	1	0	563	0	75	210	1	1	581	0	65
70	1	1	337	0	65	211	1	1	176	0	55
71	1	0	613	1	60	212	1	1	30	0	60
72	1	1	192	1	80	213	1	1	41	0	60
73	1	0	405	0	80	214	1	0	543	0	40
74	1	1	667	0	50	215	1	0	210	1	50
75	1	0	905	0	80	216	1	1	193	1	70
76	1	1	247	0	70	217	1	1	434	0	55
77	1	1	821	0	80	218	1	1	367	0	45
78	1	1	821	1	75	219	1	1	348	1	60
79	1	0	517	0	45	220	1	0	28	0	50
80	1	0	346	1	60	221	1	0	337	0	40
81	1	1	294	0	65	222	1	0	175	1	60
82	1	1	244	1	60	223	2	1	149	1	80
83	1	1	95	1	60	224	1	1	546	1	50
84	1	1	376	1	55	225	1	1	84	0	45
85	1	1	212	0	40	226	1	0	283	1	80
86	1	1	96	0	70	227	1	1	533	0	55
87	1	1	532	0	80	228	1	1	207	1	50
88	1	1	522	1	70	229	1	1	216	0	50
89	1	1	679	0	35	230	1	0	28	0	50

ID	*Clinic*	*Status*	*Time*	*Prison*	*Dose*	*ID*	*Clinic*	*Status*	*Time*	*Prison*	*Dose*
90	1	0	408	0	50	231	1	1	67	1	50
91	1	0	840	0	80	232	1	0	62	1	60
92	1	0	148	1	65	233	1	0	111	0	55
93	1	1	168	0	65	234	1	1	257	1	60
94	1	1	489	0	80	235	1	1	136	1	55
95	1	0	541	0	80	236	1	0	342	0	60
96	1	1	205	0	50	237	2	1	41	0	40
97	1	0	475	1	75	238	2	0	531	1	45
98	1	1	237	0	45	239	1	0	98	0	40
99	1	1	517	0	70	240	1	1	145	1	55
100	1	1	749	0	70	241	1	1	50	0	50
101	1	1	150	1	80	242	1	0	53	0	50
102	1	1	465	0	65	243	1	0	103	1	50
103	2	1	708	1	60	244	1	0	2	1	60
104	2	0	713	0	50	245	1	1	157	1	60
105	2	0	146	0	50	246	1	1	75	1	55
106	2	1	450	0	55	247	1	1	19	1	40
109	2	0	555	0	80	248	1	1	35	0	60
110	2	1	460	0	50	249	2	0	394	1	80
111	2	0	53	1	60	250	1	1	117	0	40
113	2	1	122	1	60	251	1	1	175	1	60
114	2	1	35	1	40	252	1	1	180	1	60
118	2	0	532	0	70	253	1	1	314	0	70
119	2	0	684	0	65	254	1	0	480	0	50
120	2	0	769	1	70	255	1	0	325	1	60
121	2	0	591	0	70	256	2	1	280	0	90
122	2	0	769	1	40	257	1	1	204	0	50
123	2	0	609	1	100	258	2	1	366	0	55
124	2	0	932	1	80	259	2	0	531	1	50
125	2	0	932	1	80	260	1	1	59	1	45
126	2	0	587	0	110	261	1	1	33	1	60
127	2	1	26	0	40	262	2	1	540	0	80
128	2	0	72	1	40	263	2	0	551	0	65
129	2	0	641	0	70	264	1	1	90	0	40
131	2	0	367	0	70	266	1	1	47	0	45

Display 12.2

12.2 Describing Survival and Cox's Regression Model

Of central importance in the analysis of survival time data are two functions used to describe their distribution, namely, the *survival function* and the *hazard function*.

12.2.1 Survival Function

Using T to denote survival time, the survival function $S(t)$ is defined as the probability that an individual survives longer than t.

$$S(t) = \Pr(T > t) \tag{12.1}$$

The graph of $S(t)$ vs. t is known as the survival curve and is useful in assessing the general characteristics of a set of survival times.

Estimating $S(t)$ from sample data is straightforward when there are no censored observations, when $\hat{S}(t)$ is simply the proportion of survival times in the sample greater than t. When, as is generally the case, the data do contain censored observations, estimation of $S(t)$ becomes more complex. The most usual estimator is now the *Kaplan-Meier* or *product limit estimator*. This involves first ordering the survival times from the smallest to the largest, $t_{(1)} \leq t_{(2)} \leq \ldots \leq t_{(n)}$, and then applying the following formula to obtain the required estimate.

$$\hat{S}(t) = \prod_{j|t_{(i)} \leq t} \left[1 - \frac{d_j}{r_j}\right] \tag{12.2}$$

where r_j is the number of individuals at risk just before $t_{(j)}$ and d_j is the number who experience the event of interest at $t_{(j)}$ (individuals censored at $t_{(j)}$ are included in r_j). The variance of the Kaplan-Meir estimator can be estimated as:

$$Var[\hat{S}(t)] = [\hat{S}(t)]^2 \sum_{j|t_{(i)} \leq t} \frac{d_j}{(r_j - d_j)} \tag{12.3}$$

Plotting estimated survival curves for different groups of observations (e.g., males and females, treatment A and treatment B) is a useful initial procedure for comparing the survival experience of the groups. More formally, the difference in survival experience can be tested by either a *log-rank test* or *Mantel-Haenszel test*. These tests essentially compare the observed number of "deaths" occurring at each particular time point with the number to be expected if the survival experience of the groups is the same. (Details of the tests are given in Hosmer and Lemeshow, 1999.)

12.2.2 Hazard Function

The hazard function *h(t)* is defined as the probability that an individual experiences the event of interest in a small time interval *s*, given that the individual has survived up to the beginning of this interval. In mathematical terms:

$$h(t) = \lim_{s \to 0} \Pr \frac{(\text{event in } (t, t+s), \text{ given survival up to } t)}{s} \quad (12.4)$$

The hazard function is also known as the *instantaneous failure rate* or *age-specific failure rate*. It is a measure of how likely an individual is to experience an event as a function of the age of the individual, and is used to assess which periods have the highest and which the lowest chance of "death" amongst those people alive at the time. In the very old, for example, there is a high risk of dying each year among those entering that stage of their life. The probability of any individual dying in their 100th year is, however, small because so few individuals live to be 100 years old.

The hazard function can also be defined in terms of the cumulative distribution and probability density function of the survival times as follows:

$$h(t) = \frac{f(t)}{1 - F(t)} = \frac{f(t)}{S(t)} \quad (12.5)$$

It then follows that:

$$h(t) = -\frac{d}{dt}\{\ln S(t)\} \quad (12.6)$$

and so

$$S(t) = \exp\{-H(t)\} \quad (12.7)$$

where *H(t)* is the integrated or cumulative hazard given by:

$$H(t) = \int_0^t h(u)du \quad (12.8)$$

The hazard function can be estimated as the proportion of individuals experiencing the event of interest in an interval per unit time, given that they have survived to the beginning of the interval; that is:

$$\hat{h}(t) = \text{Number of individuals experiencing an event in the interval beginning at time } t \div [(\text{Number of patients surviving at } t) \times (\text{Interval width})] \quad (12.9)$$

In practice, the hazard function may increase, decrease, remain constant, or indicate a more complicated process. The hazard function for deaths in humans has, for example, the "bathtub" shape shown in Display 12.3. It is relatively high immediately after birth, declines rapidly in the early years, and remains approximately constant before beginning to rise again during late middle age.

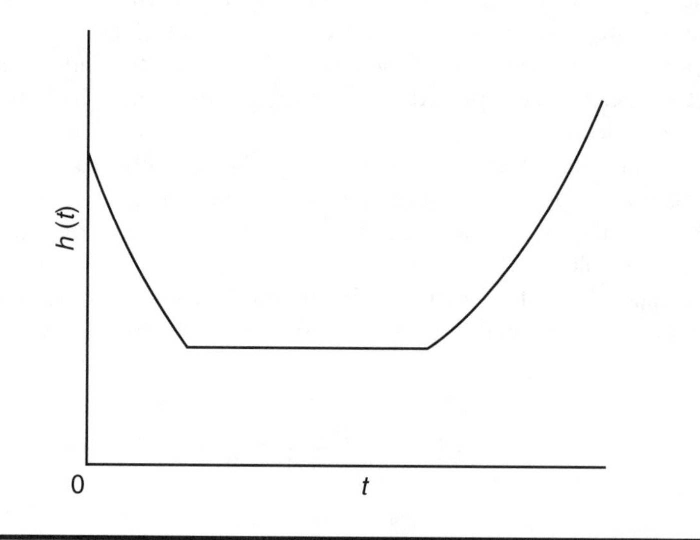

Display 12.3

12.2.3 Cox's Regression

Cox's regression is a semi-parametric approach to survival analysis in which the hazard function is modelled. The method does not require the probability distribution of the survival times to be specified; however, unlike most nonparametric methods, Cox's regression does use regression parameters in the same way as generalized linear models. The model can be written as:

$$h(t) = h_0(t)\exp(\boldsymbol{\beta}^T\boldsymbol{x}) \tag{12.10}$$

or

$$\log[h(t)] = \log[h_0(t)] + (\boldsymbol{\beta}^T\boldsymbol{x}) \tag{12.11}$$

where $\boldsymbol{\beta}$ is a vector of regression parameters and $\boldsymbol{x}$ a vector of covariate values. The hazard functions of any two individuals with covariate vectors $\boldsymbol{x}_i$ and $\boldsymbol{x}_j$ are assumed to be constant multiples of each other, the multiple being $\exp[\boldsymbol{\beta}^T(\boldsymbol{x}_i - \boldsymbol{x}_j)]$, the *hazard ratio* or incidence rate ratio. The assumption of a constant hazard ratio is called the *proportional hazards* assumption. The set of parameters $h_0(t)$ is called the *baseline hazard function*, and can be thought of as nuisance parameters whose purpose is merely to control the parameters of interest $\boldsymbol{\beta}$ for any changes in the hazard over time. The parameters $\boldsymbol{\beta}$ are estimated by maximising the *partial log-likelihood* given by:

$$\sum_j \log\left(\frac{\exp(\beta^T\mathbf{x}_f)}{\Sigma_{i \in r(f)}\exp(\beta^T\mathbf{x}_i)}\right) \tag{12.12}$$

where the first summation is over all failures f and the second summation is over all subjects $r(f)$ still alive (and therefore "at risk") at the time of failure. It can be shown that this log-likelihood is a *log profile likelihood* (i.e., the log of the likelihood in which the nuisance parameters have been replaced by functions of $\boldsymbol{\beta}$ which maximise the likelihood for fixed $\boldsymbol{\beta}$). The parameters in a Cox model are interpreted in a similar fashion to those in other regression models met in earlier chapters; that is, the estimated coefficient for an explanatory variable gives the change in the logarithm of the hazard function when the variable changes by one. A more appealing interpretation is achieved by exponentiating the coefficient, giving the effect in terms of the hazard function. An additional aid to interpretation is to calculate

$$100[\exp(\text{coefficient}) - 1] \tag{12.13}$$

The resulting value gives the percentage change in the hazard function with each unit change in the explanatory variable.

The baseline hazards can be estimated by maximising the full log-likelihood with the regression parameters evaluated at their estimated values. These hazards are nonzero only when a failure occurs. Integrating the hazard function gives the cumulative hazard function

$$H(t) = H_0(t)\exp(\boldsymbol{\beta}^T\mathbf{x}) \tag{12.14}$$

where $H_0(t)$ is the integral of $h_0(t)$. The survival curve can be obtained from $H(t)$ using Eq. (12.7).

It follows from Eq. (12.7) that the survival curve for a Cox model is given by:

$$S(t) = S_0(t)^{\exp(\boldsymbol{\beta}^T\mathbf{x})} \tag{12.15}$$

The log of the cumulative hazard function predicted by the Cox model is given by:

$$\log[H(t)] = \log H_0(t) + \boldsymbol{\beta}^T\mathbf{x} \tag{12.16}$$

so that the log cumulative hazard functions of any two subjects i and j are parallel with constant difference given by $\boldsymbol{\beta}^T(\mathbf{x}_i - \mathbf{x}_j)$.

If the subjects fall into different groups and we are not sure whether we can make the assumption that the group's hazard functions are proportional to each other, we can estimate separate log cumulative hazard functions for the groups using a stratified Cox model. These curves can then be plotted to assess whether they are sufficiently parallel. For a stratified Cox model, the partial likelihood has the same form as in Eq. (12.11) except that the risk set for a failure is not confined to subjects in the same stratum.

Survival analysis is described in more detail in Collett (1994) and in Clayton and Hills (1993).

12.3 Analysis Using SAS

12.3.1 Gastric Cancer

The data shown in Display 12.1 consist of 89 survival times. There are six values per line except the last line, which has five. The first three values belong to patients in the first treatment group and the remainder to those in the second group. The following data step constructs a suitable SAS data set.

```
data cancer;
   infile 'n:\handbook2\datasets\time.dat' expandtabs missover;
   do i = 1 to 6;
      input temp $ @;
      censor=(index(temp,'*')>0);
```

```
      temp=substr(temp,1,4);
      days=input(temp,4.);
      group=i>3;
      if days>0 then output;
   end;
   drop temp i;
run;
```

The **infile** statement gives the full path name of the file containing the ASCII data. The values are tab separated, so the **expandtabs** option is used. The **missover** option prevents SAS from going to a new line if the input statement contains more variables than there are data values, as is the case for the last line. In this case, the variable for which there is no corresponding data is set to missing.

Reading and processing the data takes place within an iterative **do** loop. The **input** statement reads one value into a character variable, **temp**. A character variable is used to allow for processing of the asterisks that indicate censored values, as there is no space between the number and the asterisk. The trailing **@** holds the line for further data to be read from it.

If **temp** contains an asterisk, the **index** function gives its position; if not, the result is zero. The **censor** variable is set accordingly. The **substr** function takes the first four characters of **temp** and the **input** function reads this into a numeric variable, **days**.

If the value of **days** is greater than zero, an observation is output to the data set. This has the effect of excluding the missing value generated because the last line only contains five values.

Finally, the character variable **temp** and the loop index variable **i** are dropped from the data set, as they are no longer needed.

With a complex data step like this, it would be wise to check the resulting data set, for example, with **proc print.**

Proc lifetest can be used to estimate and compare the survival functions of the two groups of patients as follows:

```
proc lifetest data=cancer plots=(s);
   time days*censor(1);
   strata group;
symbol1 l=1;
symbol2 l=3;
run;
```

The **plots=(s)** option on the **proc** statement specifies that survival curves be plotted. Log survival (**ls**), log-log survival (**lls**), hazard (**h**), and PDF (**p**) are other functions that may be plotted as well as a plot of censored values by strata (**c**). A list of plots can be specified; for example, **plots=(s,ls,lls)**.

The **time** statement specifies the survival time variable followed by an asterisk and the censoring variable, with the value(s) indicating a censored observation in parentheses. The censoring variable must be numeric, with non-missing values for *both* censored and uncensored observations.

The **strata** statement indicates the variable, or variables, that determine the strata levels.

Two **symbol** statements are used to specify different line types for the two groups. (The default is to use different colours, which is not very useful in black and white!)

The output is shown in Display 12.4 and the plot in Display 12.5. In Display 12.4, we find that the median survival time in group 1 is 254 with 95% confidence interval of (193, 484). In group 2, the corresponding values are 506 and (383, 676). The log-rank test for a difference in the survival curves of the two groups has an associated P-value of 0.4521. This suggests that there is no difference in the survival experience of the two groups. The likelihood ratio test (see Lawless [1982]) leads to the same conclusion, but the Wilcoxon test (see Kalbfleisch and Prentice [1980]) has an associated P-value of 0.0378, indicating that there is a difference in the survival time distributions of the two groups. The reason for the difference is that the log-rank test (and the likelihood ratio test) are most useful when the population survival curves of the two groups do not cross, indicating that the hazard functions of the two groups are proportional (see Section 12.2.3). Here the sample survival curves do cross (see Display 12.5) suggesting perhaps that the population curves might also cross. When there is a crossing of the survival curves, the Wilcoxon test is more powerful than the other tests.

```
                    The LIFETEST Procedure

                    Stratum 1: group = 0

               Product-Limit Survival Estimates

                                   Survival
                                   Standard    Number     Number
   days     Survival    Failure       Error    Failed       Left

   0.00     1.0000          0           0         0          45
  17.00     0.9778     0.0222      0.0220         1          44
  42.00     0.9556     0.0444      0.0307         2          43
  44.00     0.9333     0.0667      0.0372         3          42
  48.00     0.9111     0.0889      0.0424         4          41
```

60.00	0.8889	0.1111	0.0468	5	40
72.00	0.8667	0.1333	0.0507	6	39
74.00	0.8444	0.1556	0.0540	7	38
95.00	0.8222	0.1778	0.0570	8	37
103.00	0.8000	0.2000	0.0596	9	36
108.00	0.7778	0.2222	0.0620	10	35
122.00	0.7556	0.2444	0.0641	11	34
144.00	0.7333	0.2667	0.0659	12	33
167.00	0.7111	0.2889	0.0676	13	32
170.00	0.6889	0.3111	0.0690	14	31
183.00	0.6667	0.3333	0.0703	15	30
185.00	0.6444	0.3556	0.0714	16	29
193.00	0.6222	0.3778	0.0723	17	28
195.00	0.6000	0.4000	0.0730	18	27
197.00	0.5778	0.4222	0.0736	19	26
208.00	0.5556	0.4444	0.0741	20	25
234.00	0.5333	0.4667	0.0744	21	24
235.00	0.5111	0.4889	0.0745	22	23
254.00	0.4889	0.5111	0.0745	23	22
307.00	0.4667	0.5333	0.0744	24	21
315.00	0.4444	0.5556	0.0741	25	20
401.00	0.4222	0.5778	0.0736	26	19
445.00	0.4000	0.6000	0.0730	27	18
464.00	0.3778	0.6222	0.0723	28	17
484.00	0.3556	0.6444	0.0714	29	16
528.00	0.3333	0.6667	0.0703	30	15
542.00	0.3111	0.6889	0.0690	31	14
567.00	0.2889	0.7111	0.0676	32	13
577.00	0.2667	0.7333	0.0659	33	12
580.00	0.2444	0.7556	0.0641	34	11
795.00	0.2222	0.7778	0.0620	35	10
855.00	0.2000	0.8000	0.0596	36	9
1174.00*	.	.	.	36	8
1214.00	0.1750	0.8250	0.0572	37	7
1232.00*	.	.	.	37	6
1366.00	0.1458	0.8542	0.0546	38	5
1455.00*	.	.	.	38	4
1585.00*	.	.	.	38	3
1622.00*	.	.	.	38	2
1626.00*	.	.	.	38	1
1736.00*	.	.	.	38	0

NOTE: The marked survival times are censored observations.

Summary Statistics for Time Variable days

The LIFETEST Procedure

Quartile Estimates

Percent	Point Estimate	95% Confidence Interval (Lower)	(Upper)
75	580.00	464.00	.
50	254.00	193.00	484.00
25	144.00	74.00	195.00

Mean	Standard Error
491.84	71.01

NOTE: The mean survival time and its standard error were underestimated because the largest observation was censored and the estimation was restricted to the largest event time.

The LIFETEST Procedure

Stratum 2: group = 1

Product-Limit Survival Estimates

Days	Survival	Failure	Survival Standard Error	Number Failed	Number Left
0.00	1.0000	0	0	0	44
1.00	0.9773	0.0227	0.0225	1	43
63.00	0.9545	0.0455	0.0314	2	42
105.00	0.9318	0.0682	0.0380	3	41
125.00	0.9091	0.0909	0.0433	4	40
182.00	0.8864	0.1136	0.0478	5	39
216.00	0.8636	0.1364	0.0517	6	38
250.00	0.8409	0.1591	0.0551	7	37
262.00	0.8182	0.1818	0.0581	8	36
301.00	.	.	.	9	35
301.00	0.7727	0.2273	0.0632	10	34
342.00	0.7500	0.2500	0.0653	11	33
354.00	0.7273	0.2727	0.0671	12	32
356.00	0.7045	0.2955	0.0688	13	31
358.00	0.6818	0.3182	0.0702	14	30
380.00	0.6591	0.3409	0.0715	15	29
383.00	.	.	.	16	28

383.00	0.6136	0.3864	0.0734	17	27
388.00	0.5909	0.4091	0.0741	18	26
394.00	0.5682	0.4318	0.0747	19	25
408.00	0.5455	0.4545	0.0751	20	24
460.00	0.5227	0.4773	0.0753	21	23
489.00	0.5000	0.5000	0.0754	22	22
523.00	0.4773	0.5227	0.0753	23	21
524.00	0.4545	0.5455	0.0751	24	20
535.00	0.4318	0.5682	0.0747	25	19
562.00	0.4091	0.5909	0.0741	26	18
569.00	0.3864	0.6136	0.0734	27	17
675.00	0.3636	0.6364	0.0725	28	16
676.00	0.3409	0.6591	0.0715	29	15
748.00	0.3182	0.6818	0.0702	30	14
778.00	0.2955	0.7045	0.0688	31	13
786.00	0.2727	0.7273	0.0671	32	12
797.00	0.2500	0.7500	0.0653	33	11
955.00	0.2273	0.7727	0.0632	34	10
968.00	0.2045	0.7955	0.0608	35	9
977.00	0.1818	0.8182	0.0581	36	8
1245.00	0.1591	0.8409	0.0551	37	7
1271.00	0.1364	0.8636	0.0517	38	6
1420.00	0.1136	0.8864	0.0478	39	5
1460.00*	.	.	.	39	4
1516.00*	.	.	.	39	3
1551.00	0.0758	0.9242	0.0444	40	2
1690.00*	.	.	.	40	1
1694.00	0	1.0000	0	41	0

NOTE: The marked survival times are censored observations.

Summary Statistics for Time Variable days

Quartile Estimates

Percent	Point Estimate	95% Confidence Interval (Lower)	(Upper)
75	876.00	569.00	1271.00
50	506.00	383.00	676.00
25	348.00	250.00	388.00

The LIFETEST Procedure

Mean	Standard Error
653.22	72.35

Summary of the Number of Censored and Uncensored Values

Stratum	group	Total	Failed	Censored	Percent Censored
1	0	45	38	7	5.56
2	1	44	41	3	6.82
Total		89	79	10	11.24

The LIFETEST Procedure

Testing Homogeneity of Survival Curves for days over Strata

Rank Statistics

group	Log-Rank	Wilcoxon
0	3.3043	502.00
1	-3.3043	-502.00

Covariance Matrix for the Log-Rank Statistics

group	0	1
0	19.3099	-19.3099
1	-19.3099	19.3099

Covariance Matrix for the Wilcoxon Statistics

group	0	1
0	58385.0	-58385.0
1	-58385.0	58385.0

Test of Equality over Strata

Test	Chi-Square	DF	Pr > Chi-Square
Log-Rank	0.5654	1	0.4521
Wilcoxon	4.3162	1	0.0378
-2Log(LR)	0.3574	1	0.5500

Display 12.4

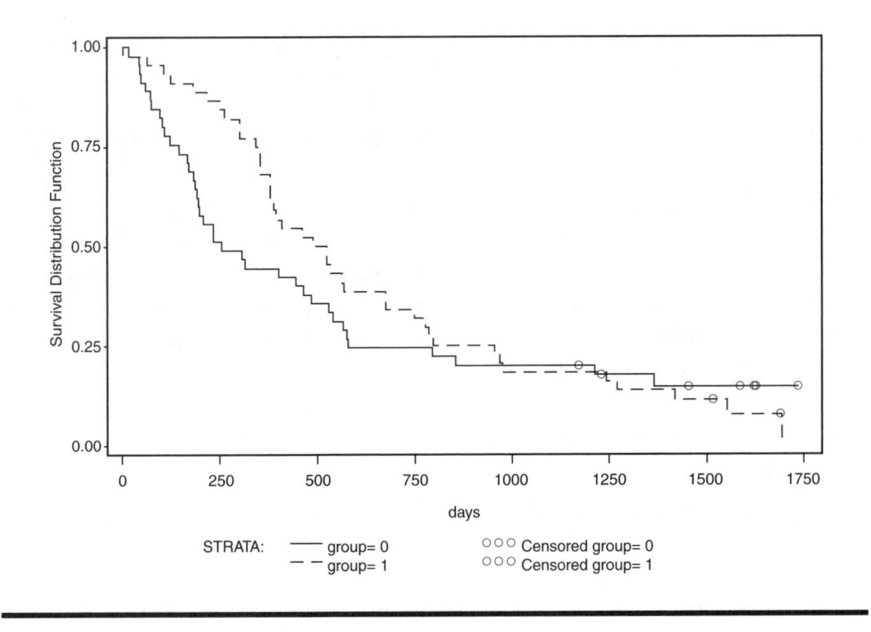

Display 12.5

12.3.2 Methadone Treatment of Heroin Addicts

The data on treatment of heroin addiction shown in Display 12.2 can be read in with the following data step.

```
data heroin;
   infile 'n:\handbook2\datasets\heroin.dat' expandtabs;
   input id clinic status time prison dose @@;
run;
```

Each line contains the data values for two observations, but there is no relevant difference between those that occur first and second. This being the case, the data can be read using list input and a double trailing @. This holds the current line for further data to be read from it. The difference between the double trailing @ and the single trailing @, used for the cancer data, is that the double @ will hold the line across iterations of the data step. SAS will only go on to a new line when it runs out of data on the current line.

The SAS log will contain the message "NOTE: SAS went to a new line when INPUT statement reached past the end of a line," which is not a cause for concern in this case. It is also worth noting that although the ID variable ranges from 1 to 266, there are actually 238 observations in the data set.

Cox regression is implemented within SAS in the phreg procedure.

The data come from two different clinics and it is possible, indeed likely, that these clinics have different hazard functions which may well not be parallel. A Cox regression model with clinics as strata and the other two variables, dose and prison, as explanatory variables can be fitted in SAS using the phreg procedure.

```
proc phreg data=heroin;
   model time*status(0)=prison dose / rl;
   strata clinic;
run;
```

In the model statement, the response variable (i.e., the failure time) is followed by an asterisk, the name of the censoring variable, and a list of censoring value(s) in parentheses. As with proc reg, the predictors must all be numeric variables. There is no built-in facility for dealing with categorical predictors, interactions, etc. These must all be calculated as separate numeric variables and dummy variables.

The rl (risklimits) option requests confidence limits for the hazard ratio. By default, these are the 95% limits.

The strata statement specifies a stratified analysis with clinics forming the strata.

The output is shown in Display 12.6. Examining the maximum likelihood estimates, we find that the parameter estimate for prison is 0.38877 and that for dose –0.03514. Interpretation becomes simpler if we concentrate on the exponentiated versions of those given under Hazard Ratio. Using the approach given in Eq. (12.13), we see first that subjects with a prison history are 47.5% more likely to complete treatment than those without a prison history. And for every increase in methadone dose by one unit (1mg), the hazard is multiplied by 0.965. This coefficient is very close to 1, but this may be because 1 mg methadone is not a large quantity. In fact, subjects in this study differ from each other by 10 to 15 units, and thus it may be more informative to find the hazard ratio of two subjects differing by a standard deviation unit. This can be done simply by rerunning the analysis with the dose standardized to zero mean and unit variance;

The PHREG Procedure

Model Information

Data Set	WORK.HEROIN
Dependent Variable	time
Censoring Variable	status
Censoring Value(s)	0
Ties Handling	BRESLOW

Summary of the Number of Event and Censored Values

Stratum	clinic	Total	Event	Censored	Percent Censored
1	1	163	122	41	25.15
2	2	75	28	47	62.67
Total		238	150	88	36.97

Convergence Status

Convergence criterion (GCONV=1E-8) satisfied.

Model Fit Statistics

Criterion	Without Covariates	With Covariates
-2 LOG L	1229.367	1195.428
AIC	1229.367	1199.428
SBC	1229.367	1205.449

Testing Global Null Hypothesis: BETA=0

Test	Chi-Square	DF	Pr > ChiSq
Likelihood Ratio	33.9393	2	<.0001
Score	33.3628	2	<.0001
Wald	32.6858	2	<.0001

Analysis of Maximum Likelihood Estimates

Variable	DF	Parameter Estimate	Standard Error	Chi-Square	Pr > ChiSq	Hazard Ratio	95% Hazard Ratio Confidence Limits	
prison	1	0.38877	0.16892	5.2974	0.0214	1.475	1.059	2.054
dose	1	-0.03514	0.00647	29.5471	<.0001	0.965	0.953	0.978

Display 12.6

The analysis can be repeated with dose standardized to zero mean and unit variance as follows:

```
proc stdize data=heroin out=heroin2;
   var dose;

proc phreg data=heroin2;
   model time*status(0)=prison dose / rl;
   strata clinic;
   baseline out=phout loglogs=lls / method=ch;

symbol1 i=join v=none l=1;
symbol2 i=join v=none l=3;
proc gplot data=phout;
   plot lls*time=clinic;
run;
```

The stdize procedure is used to standardize dose (proc standard could also have been used). Zero mean and unit variance is the default method of standardization. The resulting data set is given a different name with the out= option and the variable to be standardized is specified with the var statement.

The phreg step uses this new data set to repeat the analysis. The baseline statement is added to save the log cumulative hazards in the data set phout. loglogs=lls specifies that the log of the negative log of survival is to be computed and stored in the variable lls. The product limit estimator is the default and method=ch requests the alternative empirical cumulative hazard estimate.

Proc gplot is then used to plot the log cumulative hazard with the symbol statements defining different linetypes for each clinic.

The output from the **phreg** step is shown in Display 12.7 and the plot in Display 12.8. The coefficient of dose is now –0.50781 and the hazard ratio is 0.602. This can be interpreted as indicating a decrease in the hazard by 40% when the methadone dose increases by one standard deviation unit. Clearly, an increase in methadone dose decreases the likelihood of the addict completing treatment.

In Display 12.8, the increment at each event represents the estimated logs of the hazards at that time. Clearly, the curves are not parallel, underlying that treating the clinics as strata was sensible.

```
                         The PHREG Procedure

                          Model Information

               Data Set                  WORK.HEROIN2
               Dependent Variable        time
               Censoring Variable        status
               Censoring Value(s)        0
               Ties Handling             BRESLOW


      Summary of the Number of Event and Censored Values

                                                          Percent
  Stratum    Clinic    Total    Event     Censored      Censored

        1    1           163      122           41         25.15
        2    2            75       28           47         62.67
  -------------------------------------------------------------------
    Total                238      150           88         36.97


                          Convergence Status

          Convergence criterion (GCONV=1E-8) satisfied.

                        Model Fit Statistics

                              Without             With
               Criterion    Covariates      Covariates

               -2 LOG L       1229.367        1195.428
               AIC            1229.367        1199.428
               SBC            1229.367        1205.449
```

Testing Global Null Hypothesis: BETA=0

Test	Chi-Square	DF	Pr > ChiSq
Likelihood Ratio	33.9393	2	<.0001
Score	33.3628	2	<.0001
Wald	32.6858	2	<.0001

Analysis of Maximum Likelihood Estimates

Variable	DF	Parameter Estimate	Standard Error	Chi-Square	Pr > ChiSq	Hazard Ratio	95% Hazard Ratio Confidence Limits	
prison	1	0.38877	0.16892	5.2974	0.0214	1.475	1.059	2.054
dose	1	-0.50781	0.09342	29.5471	<.0001	0.602	0.501	0.723

Display 12.7

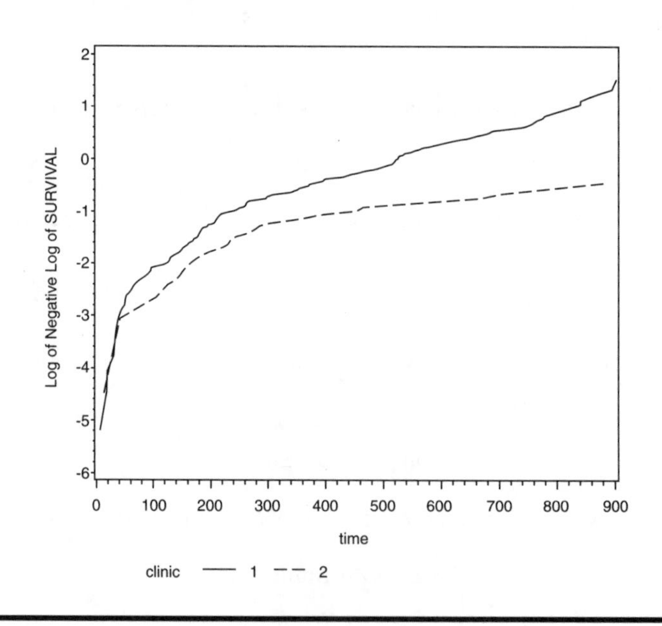

Display 12.8

Exercises

12.1 In the original analyses of the data in this chapter (see Caplehorn and Bell, 1991), it was judged that the hazards were approximately proportional for the first 450 days (see Display 12.8). Consequently, the data for this time period were analysed using clinic as a covariate rather than by stratifying on clinic. Repeat this analysis using clinic, prison, and standardized dose as covariates.

12.2 Following Caplehorn and Bell (1991), repeat the analyses in Exercise 12.1 but now treating dose as a categorical variable with three levels (<60, 60–79, ≥80) and plot the predicted survival curves for the three dose categories when prison takes the value 0 and clinic the value 1.

12.3 Test for an interaction between clinic and methadone using both continuous and categorical scales for dose.

12.4 Investigate the use of residuals in fitting a Cox regression using some of the models fitted in the text and in the previous exercises.

Chapter 13

Principal Components Analysis and Factor Analysis: The Olympic Decathlon and Statements about Pain

13.1 Description of Data

This chapter concerns two data sets: the first, given in Display 13.1 (*SDS*, Table 357), involves the results for the men's decathlon in the 1988 Olympics, and the second, shown in Display 13.2, arises from a study concerned with the development of a standardized scale to measure beliefs about controlling pain (Skevington [1990]). Here, a sample of 123 people suffering from extreme pain were asked to rate nine statements about pain on a scale of 1 to 6, ranging from disagreement to agreement. It is the correlations between these statements that appear in Display 13.2 (*SDS*, Table 492). The nine statements used were as follows:

1. Whether or not I am in pain in the future depends on the skills of the doctors.

Athlete	100 m Run	Long Jump	Shot	High Jump	400 m Run	100 m Hurdles	Discus	Pole Vault	Javelin	1500 m Run	Score
Schenk	11.25	7.43	15.48	2.27	48.90	15.13	49.28	4.7	61.32	268.95	8488
Voss	10.87	7.45	14.97	1.97	47.71	14.46	44.36	5.1	61.76	273.02	8399
Steen	11.18	7.44	14.20	1.97	48.29	14.81	43.66	5.2	64.16	263.20	8328
Thompson	10.62	7.38	15.02	2.03	49.06	14.72	44.80	4.9	64.04	285.11	8306
Blondel	11.02	7.43	12.92	1.97	47.44	14.40	41.20	5.2	57.46	256.64	8286
Plaziat	10.83	7.72	13.58	2.12	48.34	14.18	43.06	4.9	52.18	274.07	8272
Bright	11.18	7.05	14.12	2.06	49.34	14.39	41.68	5.7	61.60	291.20	8216
De wit	11.05	6.95	15.34	2.00	48.21	14.36	41.32	4.8	63.00	265.86	8189
Johnson	11.15	7.12	14.52	2.03	49.15	14.66	42.36	4.9	66.46	269.62	8180
Tarnovetsky	11.23	7.28	15.25	1.97	48.60	14.76	48.02	5.2	59.48	292.24	8167
Keskitalo	10.94	7.45	15.34	1.97	49.94	14.25	41.86	4.8	66.64	295.89	8143
Gaehwiler	11.18	7.34	14.48	1.94	49.02	15.11	42.76	4.7	65.84	256.74	8114
Szabo	11.02	7.29	12.92	2.06	48.23	14.94	39.54	5.0	56.80	257.85	8093
Smith	10.99	7.37	13.61	1.97	47.83	14.70	43.88	4.3	66.54	268.97	8083
Shirley	11.03	7.45	14.20	1.97	48.94	15.44	41.66	4.7	64.00	267.48	8036
Poelman	11.09	7.08	14.51	2.03	49.89	14.78	43.20	4.9	57.18	268.54	8021
Olander	11.46	6.75	16.07	2.00	51.28	16.06	50.66	4.8	72.60	302.42	7869
Freimuth	11.57	7.00	16.60	1.94	49.84	15.00	46.66	4.9	60.20	286.04	7860
Warming	11.07	7.04	13.41	1.94	47.97	14.96	40.38	4.5	51.50	262.41	7859
Hraban	10.89	7.07	15.84	1.79	49.68	15.38	45.32	4.9	60.48	277.84	7781
Werthner	11.52	7.36	13.93	1.94	49.99	15.64	38.82	4.6	67.04	266.42	7753
Gugler	11.49	7.02	13.80	2.03	50.60	15.22	39.08	4.7	60.92	262.93	7745
Penalver	11.38	7.08	14.31	2.00	50.24	14.97	46.34	4.4	55.68	272.68	7743
Kruger	11.30	6.97	13.23	2.15	49.98	15.38	38.72	4.6	54.34	277.84	7623
Lee Fu-An	11.00	7.23	13.15	2.03	49.73	14.96	38.06	4.5	52.82	285.57	7579
Mellado	11.33	6.83	11.63	2.06	48.37	15.39	37.52	4.6	55.42	270.07	7517
Moser	11.10	6.98	12.69	1.82	48.63	15.13	38.04	4.7	49.52	261.90	7505
Valenta	11.51	7.01	14.17	1.94	51.16	15.18	45.84	4.6	56.28	303.17	7422
O'Connell	11.26	6.90	12.41	1.88	48.24	15.61	38.02	4.4	52.68	272.06	7310
Richards	11.50	7.09	12.94	1.82	49.27	15.56	42.32	4.5	53.50	293.85	7237
Gong	11.43	6.22	13.98	1.91	51.25	15.88	46.18	4.6	57.84	294.99	7231
Miller	11.47	6.43	12.33	1.94	50.30	15.00	38.72	4.0	57.26	293.72	7016
Kwang-Ik	11.57	7.19	10.27	1.91	50.71	16.20	34.36	4.1	54.94	269.98	6907
Kunwar	12.12	5.83	9.71	1.70	52.32	17.05	27.10	2.6	39.10	281.24	5339

Display 13.1

2. Whenever I am in pain, it is usually because of something I have done or not done.
3. Whether or not I am in pain depends on what the doctors do for me.
4. I cannot get any help for my pain unless I go to seek medical advice.

5. When I am in pain, I know that it is because I have not been taking proper exercise or eating the right food.
6. People's pain results from their own carelessness.
7. I am directly responsible for my pain.
8. Relief from pain is chiefly controlled by the doctors.
9. People who are never in pain are just plain lucky.

For the decathlon data, we will investigate ways of displaying the data graphically, and, in addition, see how a statistical approach to assigning an overall score agrees with the score shown in Display 13.1, which is calculated using a series of standard conversion tables for each event.

For the pain data, the primary question of interest is: what is the underlying structure of the pain statements?

	1	2	3	4	5	6	7	8	9
1	1.0000								
2	–.0385	1.0000							
3	.6066	–.0693	1.0000						
4	.4507	–.1167	.5916	1.0000					
5	.0320	.4881	.0317	–.0802	1.0000				
6	–.2877	.4271	–.1336	–.2073	.4731	1.0000			
7	–.2974	.3045	–.2404	–.1850	.4138	.6346	1.0000		
8	.4526	–.3090	.5886	.6286	–.1397	–.1329	–.2599	1.0000	
9	.2952	–.1704	.3165	.3680	–.2367	-.1541	–.2893	.4047	1.000

Display 13.2

13.2 Principal Components and Factor Analyses

Two methods of analysis are the subject of this chapter: principal components analysis and factor analysis. In very general terms, both can be seen as approaches to summarising and uncovering any patterns in a set of multivariate data. The details behind each method are, however, quite different.

13.2.1 Principal Components Analysis

Principal components analysis is amongst the oldest and most widely used multivariate technique. Originally introduced by Pearson (1901) and independently by Hotelling (1933), the basic idea of the method is to describe the variation in a set of multivariate data in terms of a set of new, uncorrelated variables, each of which is defined to be a particular linear combination of the original variables. In other words, principal components analysis is a transformation from the observed variables, $x_1, \cdots x_p$, to variables, $y_1, \cdots, y_p$, where:

$$\begin{aligned} y_1 &= a_{11}x_1 + a_{12}x_2 + \cdots + a_{1p}x_p \\ y_2 &= a_{21}x_1 + a_{22}x_2 + \cdots + a_{2p}x_p \\ &\vdots \\ y_p &= a_{p1}x_1 + a_{p2}x_2 + \cdots + a_{pp}x_p \end{aligned} \tag{13.1}$$

The coefficients defining each new variable are chosen so that the following conditions hold:

- The y variables (the principal components) are arranged in decreasing order of variance accounted for so that, for example, the first principal component accounts for as much as possible of the variation in the original data.
- The y variables are uncorrelated with one another.

The coefficients are found as the eigenvectors of the observed covariance matrix, $\mathbf{S}$, although when the original variables are on very different scales it is wiser to extract them from the observed correlation matrix, $\mathbf{R}$, instead. The variances of the new variables are given by the eigenvectors of $\mathbf{S}$ or $\mathbf{R}$.

The usual objective of this type of analysis is to assess whether the first few components account for a large proportion of the variation in the data, in which case they can be used to provide a convenient summary of the data for later analysis. Choosing the number of components adequate for summarising a set of multivariate data is generally based on one or another of a number of relative ad hoc procedures:

- Retain just enough components to explain some specified large percentages of the total variation of the original variables. Values between 70 and 90% are usually suggested, although smaller values might be appropriate as the number of variables, p, or number of subjects, n, increases.
- Exclude those principal components whose eigenvalues are less than the average. When the components are extracted from the observed correlation matrix, this implies excluding components with eigenvalues less than 1.
- Plot the eigenvalues as a scree diagram and look for a clear "elbow" in the curve.

Principal component scores for an individual i with vector of variable values $\boldsymbol{x}_i'$ can be obtained from the equations:

$$
\begin{aligned}
y_{i1} &= \boldsymbol{a}_1'(\boldsymbol{x}_i - \bar{\boldsymbol{x}}) \\
&\vdots \\
y_{ip} &= \boldsymbol{a}_p'(\boldsymbol{x}_i - \bar{\boldsymbol{x}})
\end{aligned}
\tag{13.2}
$$

where $\boldsymbol{a}_i' = [a_{i1}, a_{i2}, \cdots, a_{ip}]$, and $\bar{\boldsymbol{x}}$ is the mean vector of the observations. (Full details of principal components analysis are given in Everitt and Dunn [2001].)

13.2.2 Factor Analysis

Factor analysis is concerned with whether the covariances or correlations between a set of observed variables can be "explained" in terms of a smaller number of unobservable latent variables or common factors. Explanation here means that the correlation between each pair of measured (manifest) variables arises because of their mutual association with the common factors. Consequently, the partial correlations between any pair of observed variables, given the values of the common factors, should be approximately zero.

The formal model linking manifest and latent variables is essentially that of multiple regression (see Chapter 3). In detail,

$$
\begin{aligned}
x_1 &= \lambda_{11}f_1 + \lambda_{12}f_2 + \cdots + \lambda_{1k}f_k + u_1 \\
x_2 &= \lambda_{21}f_1 + \lambda_{22}f_2 + \cdots + \lambda_{2k}f_k + u_2 \\
&\vdots \\
x_p &= \lambda_{p1}f_1 + \lambda_{p1}f_2 + \cdots + \lambda_{pk}f_k + u_p
\end{aligned}
\tag{13.3}
$$

where $f_1, f_2, \cdots, f_k$ are the latent variables (common factors) and $k < p$. These equations can be written more concisely as:

$$
\boldsymbol{x} = \boldsymbol{\Lambda f} + \boldsymbol{u} \tag{13.4}
$$

where

$$
\boldsymbol{\Lambda} = \begin{bmatrix} \lambda_{11} & \cdots & \lambda_{1k} \\ \vdots & & \\ \lambda_{p1} & \cdots & \lambda_{pk} \end{bmatrix}, \boldsymbol{f} = \begin{bmatrix} f_1 \\ \vdots \\ f_k \end{bmatrix}, \boldsymbol{u} = \begin{bmatrix} u_1 \\ \vdots \\ u_p \end{bmatrix}
$$

The residual terms $u_1, \cdots, u_p$, (also known as *specific variates*), are assumed uncorrelated with each other and with the common factors. The elements of Λ are usually referred to in this context as *factor loadings*.

Because the factors are unobserved, we can fix their location and scale arbitrarily. Thus, we assume they are in standardized form with mean zero and standard deviation one. (We also assume they are uncorrelated, although this is not an essential requirement.)

With these assumptions, the model in Eq. (13.4) implies that the population covariance matrix of the observed variables, $\mathbf{\Sigma}$, has the form:

$$\mathbf{\Sigma} = \mathbf{\Lambda\Lambda}' + \mathbf{\Psi} \tag{13.5}$$

where $\mathbf{\Psi}$ is a diagonal matrix containing the variances of the residual terms, $\psi_i = 1 \cdots p$.

The parameters in the factor analysis model can be estimated in a number of ways, including maximum likelihood, which also leads to a test for number of factors. The initial solution can be "rotated" as an aid to interpretation, as described fully in Everitt and Dunn (2001). (Principal components can also be rotated but then the defining maximal proportion of variance property is lost.)

13.2.3 Factor Analysis and Principal Components Compared

Factor analysis, like principal components analysis, is an attempt to explain a set of data in terms of a smaller number of dimensions than one starts with, but the procedures used to achieve this goal are essentially quite different in the two methods. Factor analysis, unlike principal components analysis, begins with a hypothesis about the covariance (or correlational) structure of the variables. Formally, this hypothesis is that a covariance matrix $\mathbf{\Sigma}$, of order and rank p, can be partitioned into two matrices $\mathbf{\Lambda\Lambda}'$ and $\mathbf{\Psi}$. The first is of order p but rank k (the number of common factors), whose off-diagonal elements are equal to those of $\mathbf{\Sigma}$. The second is a diagonal matrix of full rank p, whose elements when added to the diagonal elements of $\mathbf{\Lambda\Lambda}'$ give the diagonal elements of $\mathbf{\Sigma}$. That is, the hypothesis is that a set of k latent variables exists ($k < p$), and these are adequate to account for the interrelationships of the variables although not for their full variances. Principal components analysis, however, is merely a transformation of the data and no assumptions are made about the form of the covariance matrix from which the data arise. This type of analysis has no part corresponding to the specific variates of factor analysis. Consequently, if the factor model holds but the variances of the specific variables are small, we would expect both forms of analysis to give similar results. If, however, the specific variances are large, they will be absorbed into all the principal components, both retained and rejected; whereas factor analysis makes special provision for them. It should be remembered that both forms of analysis are similar in one important respect: namely, that they are both pointless if the observed variables are almost uncorrelated — factor analysis because it has nothing to explain and principal components analysis because it would simply lead to components that are similar to the original variables.

13.3 Analysis Using SAS

13.3.1 Olympic Decathlon

The file olympic.dat (*SDS*, p. 357) contains the event results and overall scores shown in Display 13.1, but not the athletes' names. The data are tab separated and may be read with the following data step.

```
data decathlon;
infile 'n:\handbook2\datasets\olympic.dat' expandtabs;
input run100 Ljump shot Hjump run400 hurdle discus polevlt
javelin run1500 score;
run;
```

Before undertaking a principal components analysis of the data, it is advisable to check them in some way for outliers. Here, we examine the distribution of the total score assigned to each competitor with proc univariate.

```
proc univariate data=decathlon plots;
   var score;
run;
```

Details of the proc univariate were given in Chapter 2. The output of proc univariate is given in Display 13.3.

```
                    The UNIVARIATE Procedure
                        Variable: score

                            Moments

N                          34    Sum Weights                 34
Mean               7782.85294    Sum Observations        264617
Std Deviation      594.582723    Variance            353528.614
Skewness           -2.2488675    Kurtosis            7.67309194
Uncorrected SS     2071141641    Corrected SS        11666444.3
Coeff Variation    7.63964997    Std Error Mean      101.970096
```

```
                    Basic Statistical Measures

    Location                    Variability

    Mean     7782.853     Std Deviation           594.58272
    Median   7864.500     Variance                   353529
    Mode             .    Range                        3149
                          Interquartile Range     663.00000

                   Tests for Location: Mu0=0

     Test           -Statistic-     -----p Value------

     Student's t    t   76.32486    Pr > |t|      <.0001
     Sign           M         17    Pr >= |M|     <.0001
     Signed Rank    S      297.5    Pr >= |S|     <.0001

                   Quantiles (Definition 5)

                   Quantile         Estimate

                   100% Max           8488.0
                   99%                8488.0
                   95%                8399.0
                   90%                8306.0
                   75% Q3             8180.0
                   50% Median         7864.5
                   25% Q1             7517.0
                   10%                7231.0
                   5%                 6907.0
                   1%                 5339.0
                   0% Min             5339.0

                    Extreme Observations

                 ----Lowest----    ----Highest---

                 Value     Obs     Value     Obs

                 5339       34     8286        5
                 7016       32     8328        3
                 7231       31     8399        2
                 7237       30     8488        1
```

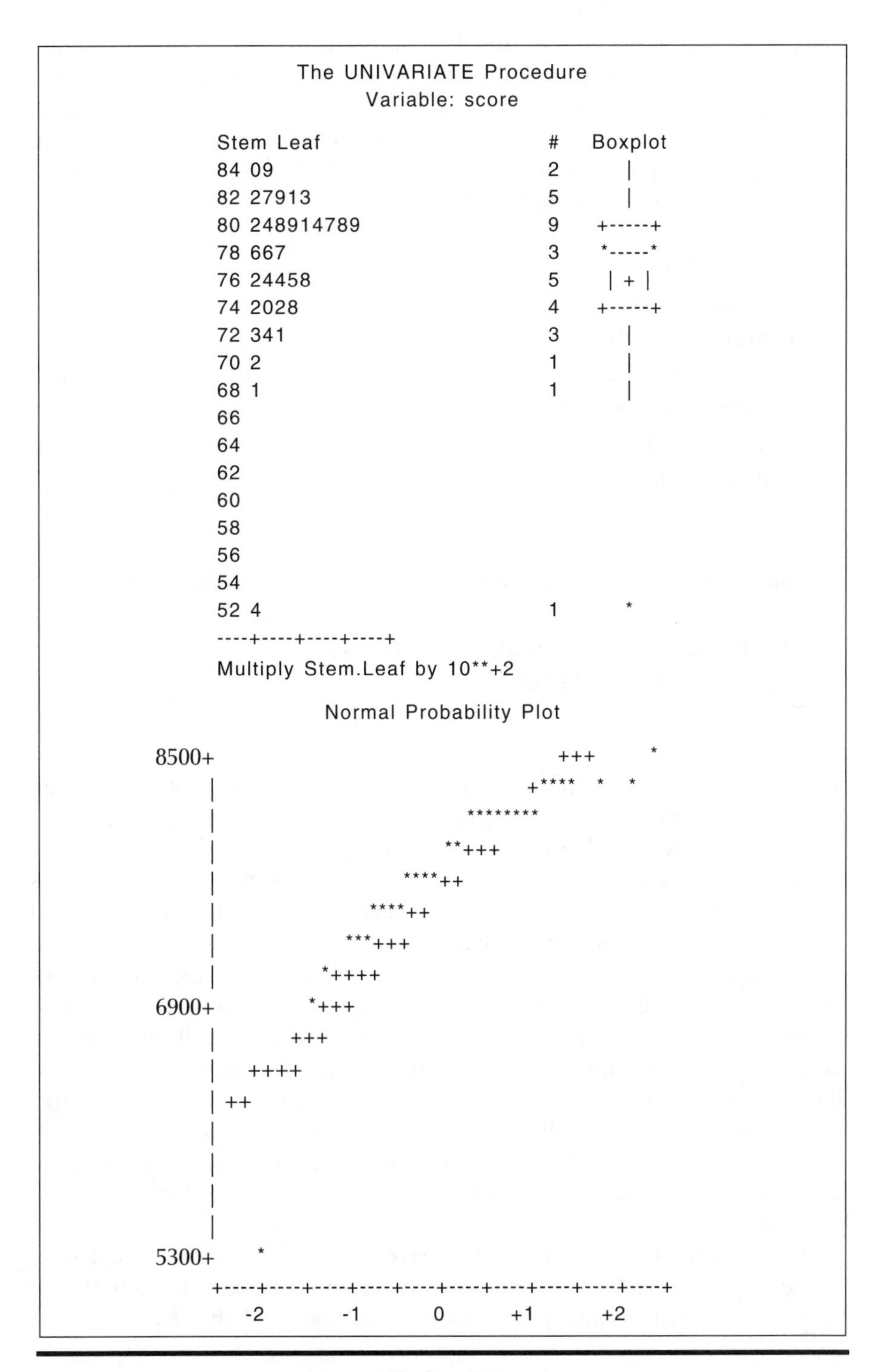

```
                The UNIVARIATE Procedure
                    Variable: score

        Stem Leaf                     #   Boxplot
        84 09                         2      |
        82 27913                      5      |
        80 248914789                  9   +-----+
        78 667                        3   *-----*
        76 24458                      5   | + |
        74 2028                       4   +-----+
        72 341                        3      |
        70 2                          1      |
        68 1                          1      |
        66
        64
        62
        60
        58
        56
        54
        52 4                          1      *
        ----+----+----+----+
        Multiply Stem.Leaf by 10**+2

                 Normal Probability Plot
  8500+                               +++      *
      |                              +****  *  *
      |                          ********
      |                        **+++
      |                     ****++
      |                  ****++
      |                ***+++
      |              *++++
  6900+             *+++
      |           +++
      |   ++++
      | ++
      |
      |
      |
      |
  5300+   *
      +----+----+----+----+----+----+----+----+----+----+
          -2        -1        0        +1        +2
```

Display 13.3

The athlete Kunwar with the lowest score is very clearly an outlier and will now be removed from the data set before further analysis. And it will help in interpreting results if all events are "scored" in the same direction; thus, we take negative values for the four running events. In this way, all ten events are such that small values represent a poor performance and large values the reverse.

```
data decathlon;
   set decathlon;
   if score > 6000;
   run100=run100*-1;
   run400=run400*-1;
   hurdle=hurdle*-1;
   run1500=run1500*-1;
run;
```

A principal components can now be applied using proc princomp:

```
proc princomp data=decathlon out=pcout;
   var run100--run1500;
run;
```

The out= option on the proc statement names a data set that will contain the principal component scores plus all the original variables. The analysis is applied to the correlation matrix by default.

The output is shown as Display 13.4. Notice first that the components as given are scaled so that the sums of squares of their elements are equal to 1. To rescale them so that they represent correlations between variables and components, they would need to be multiplied by the square root of the corresponding eigenvalue. The coefficients defining the first component are all positive and it is clearly a measure of overall performance (see later). This component has variance 3.42 and accounts for 34% of the total variation in the data. The second component contrasts performance on the "power" events such as shot and discus with the only really "stamina" event, the 1500-m run. The second component has variance 2.61; so between them, the first two components account for 60% of the total variance.

Only the first two components have eigenvalues greater than one, suggesting that the first two principal component scores for each athlete provide an adequate and parsimonious description of the data.

The PRINCOMP Procedure

Observations	33
Variables	10

Simple Statistics

	run100	Ljump	shot	Hjump	run400
Mean	-11.19636364	7.133333333	13.97636364	1.982727273	-49.27666667
StD	0.24332101	0.304340133	1.33199056	0.093983799	1.06966019

Simple Statistics

	hurdle	discus	polevlt	javelin	run1500
Mean	-15.04878788	42.35393939	4.739393939	59.43878788	-276.0384848
StD	0.50676522	3.71913123	0.334420575	5.49599841	13.6570975

Correlation Matrix

	run100	Ljump	shot	Hjump	run400	hurdle	discus	polevlt	javelin	run1500
run100	1.0000	0.5396	0.2080	0.1459	0.6059	0.6384	0.0472	0.3891	0.0647	0.2610
Ljump	0.5396	1.0000	0.1419	0.2731	0.5153	0.4780	0.0419	0.3499	0.1817	0.3956
shot	0.2080	0.1419	1.0000	0.1221	-.0946	0.2957	0.8064	0.4800	0.5977	-.2688
Hjump	0.1459	0.2731	0.1221	1.0000	0.0875	0.3067	0.1474	0.2132	0.1159	0.1141
run400	0.6059	0.5153	-.0946	0.0875	1.0000	0.5460	-.1422	0.3187	-.1204	0.5873
hurdle	0.6384	0.4780	0.2957	0.3067	0.5460	1.0000	0.1105	0.5215	0.0628	0.1433
discus	0.0472	0.0419	0.8064	0.1474	-.1422	0.1105	1.0000	0.3440	0.4429	-.4023
polevlt	0.3891	0.3499	0.4800	0.2132	0.3187	0.5215	0.3440	1.0000	0.2742	0.0315
javelin	0.0647	0.1817	0.5977	0.1159	-.1204	0.0628	0.4429	0.2742	1.0000	-.0964
run1500	0.2610	0.3956	-.2688	0.1141	0.5873	0.1433	-.4023	0.0315	-.0964	1.0000

Eigenvalues of the Correlation Matrix

	Eigenvalue	Difference	Proportion	Cumulative
1	3.41823814	0.81184501	0.3418	0.3418
2	2.60639314	1.66309673	0.2606	0.6025
3	0.94329641	0.06527516	0.0943	0.6968
4	0.87802124	0.32139459	0.0878	0.7846
5	0.55662665	0.06539914	0.0557	0.8403
6	0.49122752	0.06063230	0.0491	0.8894
7	0.43059522	0.12379709	0.0431	0.9324
8	0.30679812	0.03984871	0.0307	0.9631
9	0.26694941	0.16509526	0.0267	0.9898
10	0.10185415		0.0102	1.0000

Eigenvectors

	Prin1	Prin2	Prin3	Prin4	Prin5
run100	0.415882	-.148808	-.267472	-.088332	-.442314
Ljump	0.394051	-.152082	0.168949	0.244250	-.368914
shot	0.269106	0.483537	-.098533	0.107763	0.009755
Hjump	0.212282	0.027898	0.854987	-.387944	0.001876
run400	0.355847	-.352160	-.189496	0.080575	0.146965
hurdle	0.433482	-.069568	-.126160	-.382290	-.088803
discus	0.175792	0.503335	-.046100	-.025584	-.019359
polevlt	0.384082	0.149582	-.136872	-.143965	0.716743
javelin	0.179944	0.371957	0.192328	0.600466	-.095582
run1500	0.170143	-.420965	0.222552	0.485642	0.339772

The PRINCOMP Procedure

Eigenvectors

	Prin6	Prin7	Prin8	Prin9	Prin10
run100	-.030712	0.254398	0.663713	-.108395	0.109480
Ljump	-.093782	-.750534	-.141264	-.046139	-.055804
shot	0.230021	0.110664	-.072506	-.422476	-.650737
Hjump	0.074544	0.135124	0.155436	0.102065	-.119412
run400	0.326929	0.141339	-.146839	0.650762	-.336814
hurdle	-.210491	0.272530	-.639004	-.207239	0.259718
discus	0.614912	-.143973	-.009400	0.167241	0.534503
polevlt	-.347760	-.273266	0.276873	0.017664	0.065896
javelin	-.437444	0.341910	-.058519	0.306196	0.130932
run1500	0.300324	0.186870	0.007310	-.456882	0.243118

Display 13.4

We can use the first two principal component scores to produce a useful plot of the data, particularly if we label the points in an informative manner. This can be achieved using an annotate data set on the plot statement within proc gplot. As an example, we label the plot of the principal component scores with the athlete's overall position in the event.

```
proc rank data=pcout out=pcout descending;
  var score;
ranks posn;

data labels;
```

```
   set pcout;
   retain xsys ysys '2';
   y=prin1;
   x=prin2;
   text=put(posn,2.);
   keep xsys ysys x y text;

proc gplot data=pcout;
   plot prin1*prin2 / annotate=labels;
   symbol v=none;
run;
```

proc rank is used to calculate the finishing position in the event. The variable score is ranked in descending order and the ranks stored in the variable posn.

The annotate data set labels has variables x and y which hold the horizontal and vertical coordinates of the text to be plotted, plus the variable text which contains the label text. The two further variables that are needed, xsys and ysys, define the type of coordinate system used. A value of '2' means that the coordinate system for the annotate data set is the same as that used for the data being plotted, and this is usually what is required. As xsys and ysys are character variables, the quotes around '2' are necessary. The assignment statement text=put(posn,2.); uses the put function to convert the numeric variable posn to a character variable text that is two characters in length.

In the gplot step, the plotting symbols are suppressed by the v=none option in the symbol statement, as the aim is to plot the text defined in the annotate data set in their stead. The resulting plot is shown in Display 13.5. We comment on this plot later.

Next, we can plot the total score achieved by each athlete in the competition against each of the first two principal component scores and also find the corresponding correlations. Plots of the overall score against the first two principal components are shown in Displays 13.6 and 13.7 and the correlations in Display 13.8.

```
goptions reset=symbol;
proc gplot data=pcout;
   plot score*(prin1 prin2);
run;

proc corr data=pcout;
   var score prin1 prin2;
run;
```

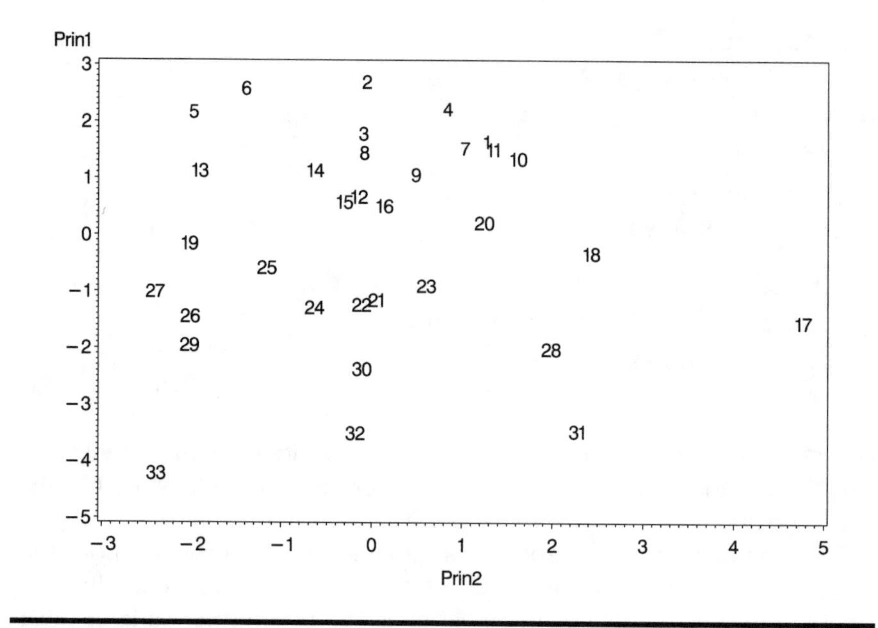

Display 13.5

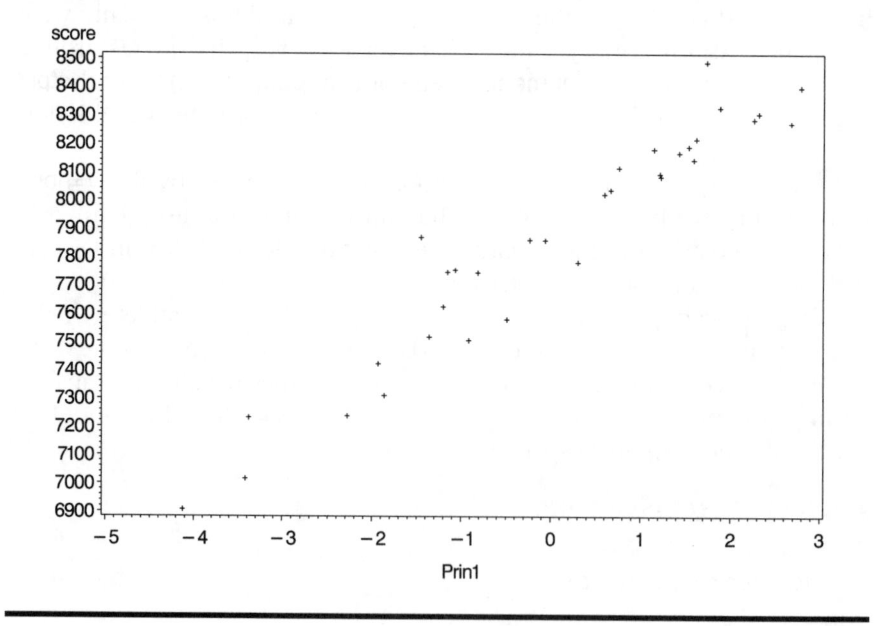

Display 13.6

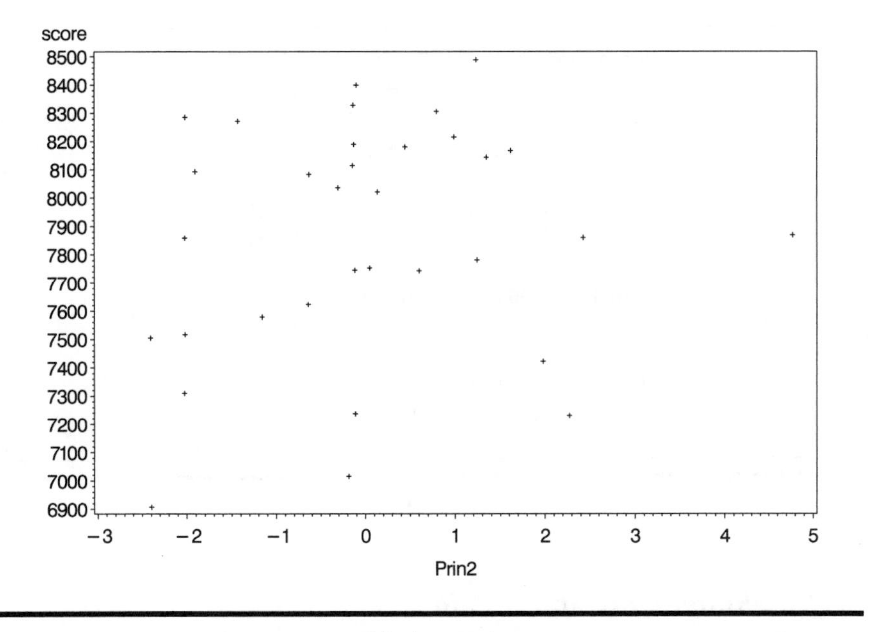

Display 13.7

Display 13.6 shows the very strong relationship between total score and first principal component score — the correlation of the two variables is found from Display 13.8 to be 0.96 which is, of course, highly significant. But the total score does not appear to be related to the second principal component score (see Display 13.7 and $r = 0.16$).

And returning to Display 13.5, the first principal component score is seen to largely rank the athletes in finishing position, confirming its interpretation as an overall measure of performance.

The CORR Procedure

3 Variables: score Prin1 Prin2

Simple Statistics

Variable	N	Mean	Std Dev	Sum	Minimum	Maximum
score	33	7857	415.06945	259278	6907	8488
Prin1	33	0	1.84885	0	-4.12390	2.78724
Prin2	33	0	1.61443	0	-2.40788	4.77144

```
        Pearson Correlation Coefficients, N = 33
              Prob > |r| under H0: Rho=0

                 score        Prin1        Prin2

   score       1.00000      0.96158      0.16194
                             <.0001       0.3679

   Prin1       0.96158      1.00000      0.00000
                <.0001                    1.0000

   Prin2       0.16194      0.00000      1.00000
                0.3679       1.0000
```

Display 13.8

13.3.2 Statements about Pain

The SAS procedure proc factor can accept data in the form of a correlation, or covariance matrix, as well as in the normal rectangular data matrix. To analyse a correlation or covariance matrix, the data need to be read into a special SAS data set with type=corr or type=cov. The correlation matrix shown in Display 13.2 was edited into the form shown in Display 13.9 and read in as follows:

```
data pain (type = corr);
infile 'n:\handbook2\datasets\pain.dat' expandtabs missover;
input _type_ $ _name_ $ p1 - p9;
run;
```

The type=corr option on the data statement specifies the type of SAS data set being created. The value of the _type_ variable indicates what type of information the observation holds. When _type_=CORR, the values of the variables are correlation coefficients. When _type_=N, the values are the sample sizes. Only the correlations are necessary but the sample sizes have been entered because they will be used by the maximum likelihood method for the test of the number of factors. The _name_ variable identifies the variable whose correlations are in that row of the matrix. The missover option in the infile statement obviates the need to enter the data for the upper triangle of the correlation matrix.

```
CORR p1    1.0
CORR p2 -.0385    1.0
CORR p3  .6066 -.0693    1.0
CORR p4  .4507 -.1167  .5916    1.0
CORR p5  .0320  .4881  .0317 -.0802    1.0
CORR p6 -.2877  .4271 -.1336 -.2073  .4731    1.0
CORR p7 -.2974  .3045 -.2404 -.1850  .4138  .6346    1.0
CORR p8  .4526 -.3090  .5886  .6286 -.1397 -.1329 -.2599    1.0
CORR p9  .2952 -.1704  .3165  .3680 -.2367 -.1541 -.2893 .4047  1.0
N    N     123     123     123     123     123     123     123    123 123
```

Display 13.9

Both principal components analysis and maximum likelihood factor analysis might be applied to the pain statement data using proc factor. The following, however, specifies a maximum likelihood factor analysis extracting two factors and requesting a scree plot, often useful in selecting the appropriate number of components. The output is shown in Display 13.10.

```
proc factor data=pain method=ml n=2 scree;
  var p1-p9;
run;
```

```
                    The FACTOR Procedure
         Initial Factor Method: Maximum Likelihood

            Prior Communality Estimates: SMC

      p1            p2            p3            p4            p5

0.46369858    0.37626982    0.54528471    0.51155233    0.39616724


             p6            p7            p8            p9

     0.55718109    0.48259656    0.56935053    0.25371373
```

Preliminary Eigenvalues: Total = 8.2234784 Average = 0.91371982

	Eigenvalue	Difference	Proportion	Cumulative
1	5.85376325	3.10928282	0.7118	0.7118
2	2.74448043	1.96962348	0.3337	1.0456
3	0.77485695	0.65957907	0.0942	1.1398
4	0.11527788	0.13455152	0.0140	1.1538
5	-.01927364	0.13309824	-0.0023	1.1515
6	-.15237189	0.07592411	-0.0185	1.1329
7	-.22829600	0.10648720	-0.0278	1.1052
8	-.33478320	0.19539217	-0.0407	1.0645
9	-.53017537		-0.0645	1.0000

2 factors will be retained by the NFACTOR criterion.

The FACTOR Procedure
Initial Factor Method: Maximum Likelihood

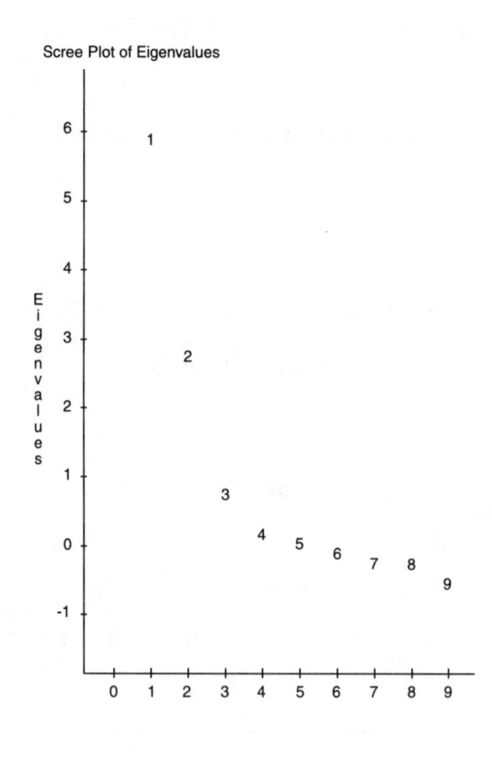

Number

The FACTOR Procedure
Initial Factor Method: Maximum Likelihood

```
Iteration   Criterion   Ridge  Change                                     Communalities

    1     0.5068620  0.0000  0.1321  0.43833  0.31605  0.67742  0.56410  0.46025  0.61806  0.51629
                                     0.57225  0.24191
    2     0.5050936  0.0000  0.0228  0.45691  0.31638  0.67459  0.55648  0.43749  0.62559  0.53374
                                     0.56905  0.23600
    3     0.5047129  0.0000  0.0115  0.45716  0.30839  0.67721  0.55492  0.43109  0.63711  0.53678
                                     0.56677  0.23488
    4     0.5046043  0.0000  0.0050  0.45798  0.30536  0.67706  0.55454  0.42611  0.64126  0.54067
                                     0.56634  0.23385
    5     0.5045713  0.0000  0.0030  0.45800  0.30332  0.67687  0.55454  0.42357  0.64424  0.54202
                                     0.56637  0.23354
    6     0.5045610  0.0000  0.0015  0.45795  0.30229  0.67662  0.55464  0.42206  0.64569  0.54297
                                     0.56649  0.23335
    7     0.5045578  0.0000  0.0009  0.45787  0.30169  0.67644  0.55473  0.42124  0.64656  0.54342
                                     0.56661  0.23327

Convergence criterion satisfied.


               Significance Tests Based on 123 Observations

                                                                  Pr >
                Test                        DF    Chi-Square     ChiSq

        H0: No common factors               36      400.8045    <.0001
        HA: At least one common factor
        H0: 2 Factors are sufficient        19       58.9492    <.0001
        HA: More factors are needed


          Chi-Square without Bartlett's Correction       61.556052
          Akaike's Information Criterion                 23.556052
          Schwarz's Bayesian Criterion                  -29.875451
          Tucker and Lewis's Reliability Coefficient      0.792510


                     Squared Canonical Correlations

                        Factor1           Factor2

                        0.86999919     0.76633961
```

Eigenvalues of the Weighted Reduced Correlation Matrix: Total = 9.97197304
Average = 1.107997

	Eigenvalue	Difference	Proportion	Cumulative
1	6.69225960	3.41254398	0.6711	0.6711
2	3.27971562	2.44878855	0.3289	1.0000
3	0.83092707	0.68869546	0.0833	1.0833
4	0.14223161	0.07800630	0.0143	1.0976
5	0.06422532	0.16781356	0.0064	1.1040
6	-.10358825	0.01730569	-0.0104	1.0936
7	-.12089394	0.19579997	-0.0121	1.0815
8	-.31669391	0.17951618	-0.0318	1.0498
9	-.49621009		-0.0498	1.0000

The FACTOR Procedure
Initial Factor Method: Maximum Likelihood

Factor Pattern

	Factor1	Factor2
p1	0.64317	0.21013
p2	-0.36053	0.41411
p3	0.71853	0.40014
p4	0.68748	0.28658
p5	-0.31016	0.56992
p6	-0.51975	0.61363
p7	-0.56193	0.47734
p8	0.70918	0.25240
p9	0.48230	0.02495

Variance Explained by Each Factor

Factor	Weighted	Unweighted
Factor1	6.69225960	2.95023735
Factor2	3.27971562	1.45141080

```
Final Communality Estimates and Variable Weights
Total Communality: Weighted = 9.971975 Unweighted = 4.401648

Variable    Communality      Weight

p1           0.45782727   1.84458678
p2           0.30146582   1.43202409
p3           0.67639995   3.09059720
p4           0.55475992   2.24582480
p5           0.42100442   1.72782049
p6           0.64669210   2.82929965
p7           0.54361175   2.19019950
p8           0.56664692   2.30737576
p9           0.23324000   1.30424476
```

Display 13.10

Here, the scree plot suggests perhaps three factors, and the formal significance test for number of factors given in Display 13.10 confirms that more than two factors are needed to adequately describe the observed correlations. Consequently, the analysis is now extended to three factors, with a request for a varimax rotation of the solution.

```
proc factor data=pain method=ml n=3 rotate=varimax;
  var p1-p9;
run;
```

The output is shown in Display 13.11. First, the test for number factors indicates that a three-factor solution provides an adequate description of the observed correlations. We can try to identify the three common factors by examining the rotated loading in Display 13.11. The first factor loads highly on statements 1, 3, 4, and 8. These statements attribute pain relief to the control of doctors, and thus we might label the factor *doctors' control of pain*. The second factor has its highest loadings on statements 6 and 7. These statements associated the cause of pain as one's own actions, and the factor might be labelled *individual's responsibility for pain*. The third factor has high loadings on statements 2 and 5. Again, both involve an individual's own responsibility for their pain but now specifically because of things they have not done; the factor might be labelled *lifestyle responsibility for pain*.

The FACTOR Procedure
Initial Factor Method: Maximum Likelihood

Prior Communality Estimates: SMC

p1	p2	p3	p4	p5
0.46369858	0.37626982	0.54528471	0.51155233	0.39616724

p6	p7	p8	p9
0.55718109	0.48259656	0.56935053	0.25371373

Preliminary Eigenvalues: Total = 8.2234784 Average = 0.91371982

	Eigenvalue	Difference	Proportion	Cumulative
1	5.85376325	3.10928282	0.7118	0.7118
2	2.74448043	1.96962348	0.3337	1.0456
3	0.77485695	0.65957907	0.0942	1.1398
4	0.11527788	0.13455152	0.0140	1.1538
5	-.01927364	0.13309824	-0.0023	1.1515
6	-.15237189	0.07592411	-0.0185	1.1329
7	-.22829600	0.10648720	-0.0278	1.1052
8	-.33478320	0.19539217	-0.0407	1.0645
9	-.53017537		-0.0645	1.0000

3 factors will be retained by the NFACTOR criterion.

Iteration	Criterion	Ridge	Change	Communalities						
1	0.1604994	0.0000	0.2170	0.58801	0.43948	0.66717	0.54503	0.55113	0.77414	0.52219
				0.75509	0.24867					
2	0.1568974	0.0000	0.0395	0.59600	0.47441	0.66148	0.54755	0.51168	0.81079	0.51814
				0.75399	0.25112					
3	0.1566307	0.0000	0.0106	0.59203	0.47446	0.66187	0.54472	0.50931	0.82135	0.51377
				0.76242	0.24803					
4	0.1566095	0.0000	0.0029	0.59192	0.47705	0.66102	0.54547	0.50638	0.82420	0.51280
				0.76228	0.24757					
5	0.1566078	0.0000	0.0008	0.59151	0.47710	0.66101	0.54531	0.50612	0.82500	0.51242
				0.76293	0.24736					

Convergence criterion satisfied.

Significance Tests Based on 123 Observations

Test	DF	Chi-Square	Pr > ChiSq
H0: No common factors	36	400.8045	<.0001
HA: At least one common factor			
H0: 3 Factors are sufficient	12	18.1926	0.1100
HA: More factors are needed			

Chi-Square without Bartlett's Correction	19.106147
Akaike's Information Criterion	-4.893853
Schwarz's Bayesian Criterion	-38.640066
Tucker and Lewis's Reliability Coefficient	0.949075

The FACTOR Procedure
Initial Factor Method: Maximum Likelihood

Squared Canonical Correlations

Factor1	Factor2	Factor3
0.90182207	0.83618918	0.60884385

Eigenvalues of the Weighted Reduced Correlation Matrix: Total = 15.8467138
Average = 1.76074598

	Eigenvalue	Difference	Proportion	Cumulative
1	9.18558880	4.08098588	0.5797	0.5797
2	5.10460292	3.54807912	0.3221	0.9018
3	1.55652380	1.26852906	0.0982	1.0000
4	0.28799474	0.10938119	0.0182	1.0182
5	0.17861354	0.08976744	0.0113	1.0294
6	0.08884610	0.10414259	0.0056	1.0351
7	-.01529648	0.16841933	-0.0010	1.0341
8	-.18371581	0.17272798	-0.0116	1.0225
9	-.35644379		-0.0225	1.0000

Factor Pattern

	Factor1	Factor2	Factor3
p1	0.60516	0.29433	0.37238
p2	-0.45459	0.29155	0.43073
p3	0.61386	0.49738	0.19172
p4	0.62154	0.39877	-0.00365
p5	-0.40635	0.45042	0.37154
p6	-0.67089	0.59389	-0.14907
p7	-0.62525	0.34279	-0.06302
p8	0.68098	0.47418	-0.27269
p9	0.44944	0.16166	-0.13855

Variance Explained by Each Factor

Factor	Weighted	Unweighted
Factor1	9.18558880	3.00788644
Factor2	5.10460292	1.50211187
Factor3	1.55652380	0.61874873

Final Communality Estimates and Variable Weights
Total Communality: Weighted = 15.846716 Unweighted = 5.128747

Variable	Communality	Weight
p1	0.59151181	2.44807030
p2	0.47717797	1.91240023
p3	0.66097328	2.94991222
p4	0.54534606	2.19927836
p5	0.50603810	2.02479887
p6	0.82501333	5.71444465
p7	0.51242072	2.05095025
p8	0.76294154	4.21819901
p9	0.24732424	1.32865993

The FACTOR Procedure
Rotation Method: Varimax

Orthogonal Transformation Matrix

	1	2	3
1	0.72941	-0.56183	-0.39027
2	0.68374	0.61659	0.39028
3	0.02137	-0.55151	0.83389

```
                Rotated Factor Pattern

               Factor1     Factor2     Factor3

        p1     0.65061    -0.36388     0.18922
        p2    -0.12303     0.19762     0.65038
        p3     0.79194    -0.14394     0.11442
        p4     0.72594    -0.10131    -0.08998
        p5     0.01951     0.30112     0.64419
        p6    -0.08648     0.82532     0.36929
        p7    -0.22303     0.59741     0.32525
        p8     0.81511     0.06018    -0.30809
        p9     0.43540    -0.07642    -0.22784

          Variance Explained by Each Factor

        Factor          Weighted    Unweighted

        Factor1       7.27423715    2.50415379
        Factor2       5.31355675    1.34062697
        Factor3       3.25892162    1.28396628

     Final Communality Estimates and Variable Weights
Total Communality: Weighted = 15.846716 Unweighted = 5.128747

        Variable    Communality        Weight

        p1           0.59151181    2.44807030
        p2           0.47717797    1.91240023
        p3           0.66097328    2.94991222
        p4           0.54534606    2.19927836
        p5           0.50603810    2.02479887
        p6           0.82501333    5.71444465
        p7           0.51242072    2.05095025
        p8           0.76294154    4.21819901
        p9           0.24732424    1.32865993
```

Display 13.11

Exercises

13.1 Repeat the principal components analysis of the Olympic decathlon data without removing the athlete who finished last in the competition. How do the results compare with those reported in this chapter (Display 13.5)?

13.2 Run a principal components analysis on the pain data and compare the results with those from the maximum likelihood factor analysis.

13.3 Run principal factor analysis and maximum likelihood factor analysis on the Olympic decathlon data. Investigate the use of other methods of rotation than varimax.

Chapter 14

Cluster Analysis: Air Pollution in the U.S.A.

14.1 Description of Data

The data to be analysed in this chapter relate to air pollution in 41 U.S. cities. The data are given in Display 14.1 (they also appear in *SDS* as Table 26). Seven variables are recorded for each of the cities:

1. SO_2 content of air, in micrograms per cubic metre
2. Average annual temperature, in °F
3. Number of manufacturing enterprises employing 20 or more workers
4. Population size (1970 census), in thousands
5. Average annual wind speed, in miles per hour
6. Average annual precipitation, in inches
7. Average number of days per year with precipitation

In this chapter we use variables 2 to 7 in a cluster analysis of the data to investigate whether there is any evidence of distinct groups of cities. The resulting clusters are then assessed in terms of their air pollution levels as measured by SO_2 content.

	1	*2*	*3*	*4*	*5*	*6*	*7*
Phoenix	10	70.3	213	582	6.0	7.05	36
Little Rock	13	61.0	91	132	8.2	48.52	100
San Francisco	12	56.7	453	716	8.7	20.66	67
Denver	17	51.9	454	515	9.0	12.95	86
Hartford	56	49.1	412	158	9.0	43.37	127
Wilmington	36	54.0	80	80	9.0	40.25	114
Washington	29	57.3	434	757	9.3	38.89	111
Jacksonville	14	68.4	136	529	8.8	54.47	116
Miami	10	75.5	207	335	9.0	59.80	128
Atlanta	24	61.5	368	497	9.1	48.34	115
Chicago	110	50.6	3344	3369	10.4	34.44	122
Indianapolis	28	52.3	361	746	9.7	38.74	121
Des Moines	17	49.0	104	201	11.2	30.85	103
Wichita	8	56.6	125	277	12.7	30.58	82
Louisville	30	55.6	291	593	8.3	43.11	123
New Orleans	9	68.3	204	361	8.4	56.77	113
Baltimore	47	55.0	625	905	9.6	41.31	111
Detroit	35	49.9	1064	1513	10.1	30.96	129
Minneapolis	29	43.5	699	744	10.6	25.94	137
Kansas City	14	54.5	381	507	10.0	37.00	99
St. Louis	56	55.9	775	622	9.5	35.89	105
Omaha	14	51.5	181	347	10.9	30.18	98
Albuquerque	11	56.8	46	244	8.9	7.77	58
Albany	46	47.6	44	116	8.8	33.36	135
Buffalo	11	47.1	391	463	12.4	36.11	166
Cincinnati	23	54.0	462	453	7.1	39.04	132
Cleveland	65	49.7	1007	751	10.9	34.99	155
Columbus	26	51.5	266	540	8.6	37.01	134
Philadelphia	69	54.6	1692	1950	9.6	39.93	115
Pittsburgh	61	50.4	347	520	9.4	36.22	147
Providence	94	50.0	343	179	10.6	42.75	125
Memphis	10	61.6	337	624	9.2	49.10	105
Nashville	18	59.4	275	448	7.9	46.00	119
Dallas	9	66.2	641	844	10.9	35.94	78
Houston	10	68.9	721	1233	10.8	48.19	103
Salt Lake City	28	51.0	137	176	8.7	15.17	89
Norfolk	31	59.3	96	308	10.6	44.68	116
Richmond	26	57.8	197	299	7.6	42.59	115

	1	*2*	*3*	*4*	*5*	*6*	*7*
Seattle	29	51.1	379	531	9.4	38.79	164
Charleston	31	55.2	35	71	6.5	40.75	148
Milwaukee	16	45.7	569	717	11.8	29.07	123

Display 14.1

14.2 Cluster Analysis

Cluster analysis is a generic term for a large number of techniques that have the common aim of determining whether a (usually) multivariate data set contains distinct groups or clusters of observations and, if so, find which of the observations belong in the same cluster. A detailed account of what is now a very large area is given in Everitt, Landau, and Leese (2001).

The most commonly used classes of clustering methods are those that lead to a series of nested or hierarchical classifications of the observations, beginning at the stage where each observation is regarded as forming a single-member "cluster" and ending at the stage where all the observations are in a single group. The complete hierarchy of solutions can be displayed as a tree diagram known as a *dendrogram*. In practice, most users are interested in choosing a particular partition of the data, that is, a particular number of groups that is optimal in some sense. This entails "cutting" the dendrogram at some particular level.

Most hierarchical methods operate not on the raw data, but on an inter-individual distance matrix calculated from the raw data. The most commonly used distance measure is Euclidean and is defined as:

$$d_{ij} = \sqrt{\sum_{k=1}^{p} (x_{ik} - x_{jk})^2} \tag{14.1}$$

where x_{ik} and x_{jk} are the values of the kth variable for observations i and j.

The different members of the class of hierarchical clustering techniques arise because of the variety of ways in which the distance between a cluster containing several observations and a single observation, or between two clusters, can be defined. The inter-cluster distances used by three commonly applied hierarchical clustering techniques are

- *Single linkage clustering*: distance between their closest observations
- *Complete linkage clustering*: distance between the most remote observations
- *Average linkage clustering*: average of distances between all pairs of observations, where members of a pair are in different groups

Important issues that often need to be considered when using clustering in practice include how to scale the variables before calculating the distance matrix, which particular method of cluster analysis to use, and how to decide on the appropriate number of groups in the data. These and many other practical problems of clustering are discussed in Everitt et al. (2001).

14.3 Analysis Using SAS

The data set for Table 26 in *SDS* does not contain the city names shown in Display 14.1; thus, we have edited the data set so that they occupy the first 16 columns. The resulting data set can be read in as follows:

```
data usair;
   infile 'n:\handbook2\datasets\usair.dat' expandtabs;
   input city $16. so2 temperature factories population wind-
speed rain rainydays;
run;
```

The names of the cities are read into the variable city with a $16. format because several of them contain spaces and are longer than the default length of eight characters. The numeric data are read in with list input.

We begin by examining the distributions of the six variables to be used in the cluster analysis.

```
proc univariate data=usair plots;
   var temperature--rainydays;
   id city;
run;
```

The univariate procedure was described in Chapter 2. Here, we use the plots option, which has the effect of including stem and leaf plots, box plots, and normal probability plots in the printed output. The id statement has the effect of labeling the extreme observations by name rather than simply by observation number.

The output for **factories** and **population** is shown in Display 14.2. Chicago is clearly an outlier, both in terms of manufacturing enterprises and population size. Although less extreme, Phoenix has the lowest value on all three climate variables (relevant output not given to save space). Both will therefore be excluded from the data set to be analysed.

```
data usair2;
   set usair;
   if city not in('Chicago','Phoenix');
run;
```

```
                    The UNIVARIATE Procedure
                     Variable:  factories

                            Moments

N                          41    Sum Weights                 41
Mean               463.097561    Sum Observations          8987
Std Deviation      563.473948    Variance             317502.89
Skewness           3.75488343    Kurtosis             17.403406
Uncorrected SS       21492949    Corrected SS        12700115.6
Coeff Variation     21.674998    Std Error Mean      87.9998462


                   Basic Statistical Measures

            Location                  Variability

        Mean     463.0976     Std Deviation          563.47395
        Median   347.0000     Variance                  317503
        Mode       .          Range                       3309
                              Interquartile Range    281.00000


                  Tests for Location: Mu0=0

         Test              -Statistic-       -----P-value------

         Student's t   t    5.262481    Pr > |t|      <.0001
         Sign          M        20.5    Pr >= |M|     <.0001
         Signed Rank   S       430.5    Pr >= |S|     <.0001
```

```
                    Quantiles (Definition 5)

                    Quantile         Estimate

                    100% Max             3344
                    99%                  3344
                    95%                  1064
                    90%                   775
                    75% Q3                462
                    50% Median            347
                    25% Q1                181
                    10%                    91
                    5%                     46
                    1%                     35
                    0% Min                 35

                     Extreme Observations

     --------------Lowest------------     ------------Highest------------

     Value   city               Obs     Value   city               Obs

        35   Charleston          40       775   St. Louis           21
        44   Albany              24      1007   Cleveland           27
        46   Albuquerque         23      1064   Detroit             18
        80   Wilmington           6      1692   Philadelphia        29
        91   Little Rock          2      3344   Chicago             11

                  The UNIVARIATE Procedure
                    Variable:  factories

          Stem Leaf                          #    Boxplot
          32 4                               1       *
          30
          28
          26
          24
          22
          20
          18
          16 9                               1       *
          14
          12
          10 16                              2       0
           8
           6 24028                           5       |
           4 135567                          6    +--+--+
           2 001178944567889                15    *-----*
           0 44589002448                    11    +-----+
            ----+----+----+----+
          Multiply Stem.Leaf by 10**+2
```

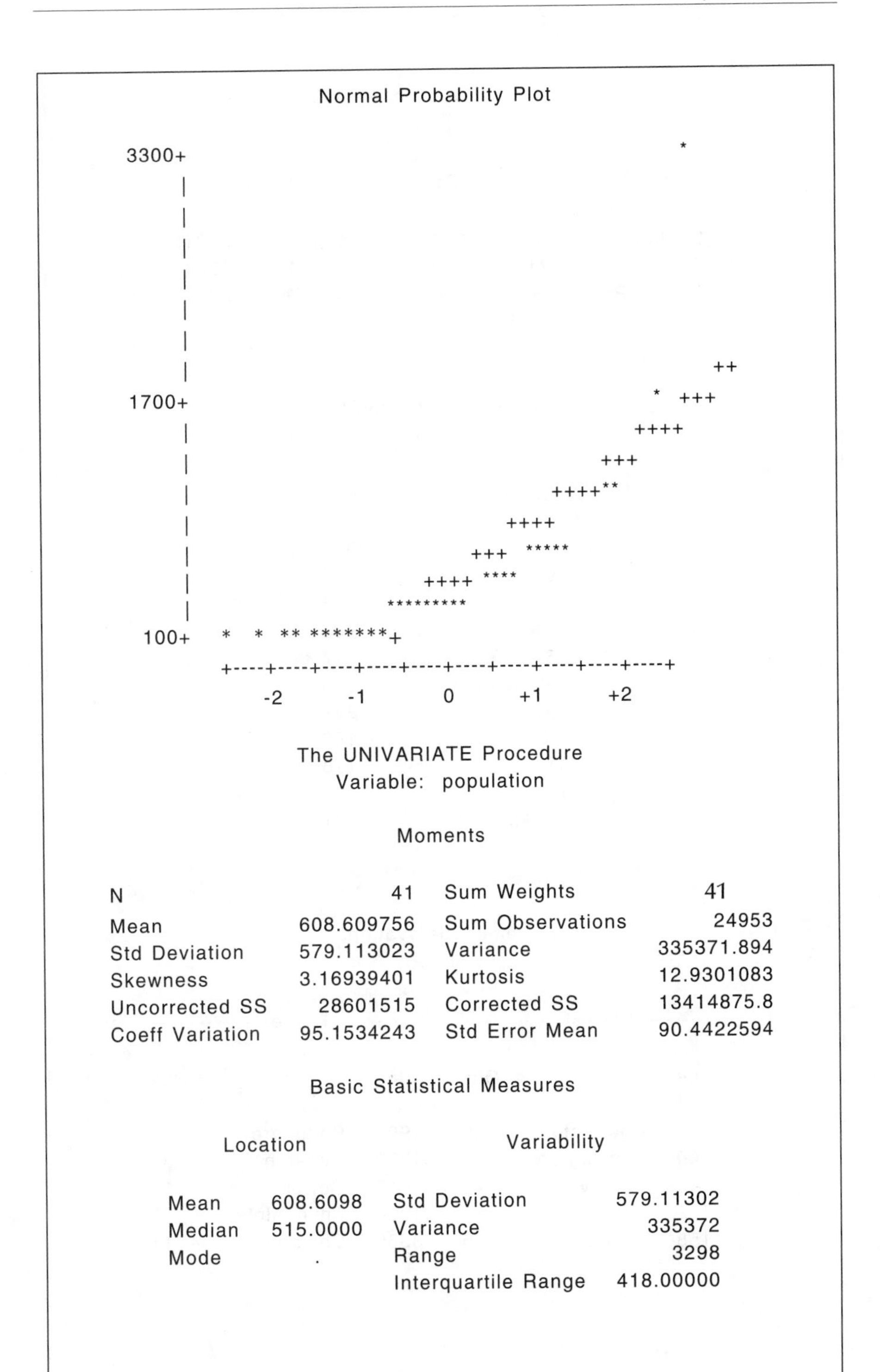

Normal Probability Plot

```
3300+                                                              *
    |
    |
    |
    |
    |
    |
    |                                                                   ++
1700+                                                           *  +++
    |                                                          ++++
    |                                                       +++
    |                                                  ++++**
    |                                             ++++
    |                                         +++  *****
    |                                    ++++ ****
    |                                *********
 100+  *  *  ** *******+
        +----+----+----+----+----+----+----+----+----+----+
            -2        -1         0        +1        +2
```

The UNIVARIATE Procedure
Variable: population

Moments

N	41	Sum Weights	41
Mean	608.609756	Sum Observations	24953
Std Deviation	579.113023	Variance	335371.894
Skewness	3.16939401	Kurtosis	12.9301083
Uncorrected SS	28601515	Corrected SS	13414875.8
Coeff Variation	95.1534243	Std Error Mean	90.4422594

Basic Statistical Measures

Location		Variability	
Mean	608.6098	Std Deviation	579.11302
Median	515.0000	Variance	335372
Mode	.	Range	3298
		Interquartile Range	418.00000

```
                  Tests for Location: Mu0=0

      Test            -Statistic-        -----P-value------

      Student's t    t   6.729263    Pr > |t|       <.0001
      Sign           M       20.5    Pr >= |M|      <.0001
      Signed Rank    S      430.5    Pr >= |S|      <.0001


                  Quantiles (Definition 5)

                  Quantile          Estimate

                  00% Max               3369
                  99%                   3369
                  95%                   1513
                  90%                    905
                  75% Q3                 717
                  50% Median             515
                  25% Q1                 299
                  0%                     158
                  5%                     116
                  1%                      71
                  0% Min                  71


                  Extreme Observations

 -------------Lowest------------   ------------Highest------------

 Value   city              Obs     Value   city              Obs

    71   Charleston         40       905   Baltimore          17
    80   Wilmington          6      1233   Houston            35
   116   Albany             24      1513   Detroit            18
   132   Little Rock         2      1950   Philadelphia       29
   158   Hartford            5      3369   Chicago            11
```

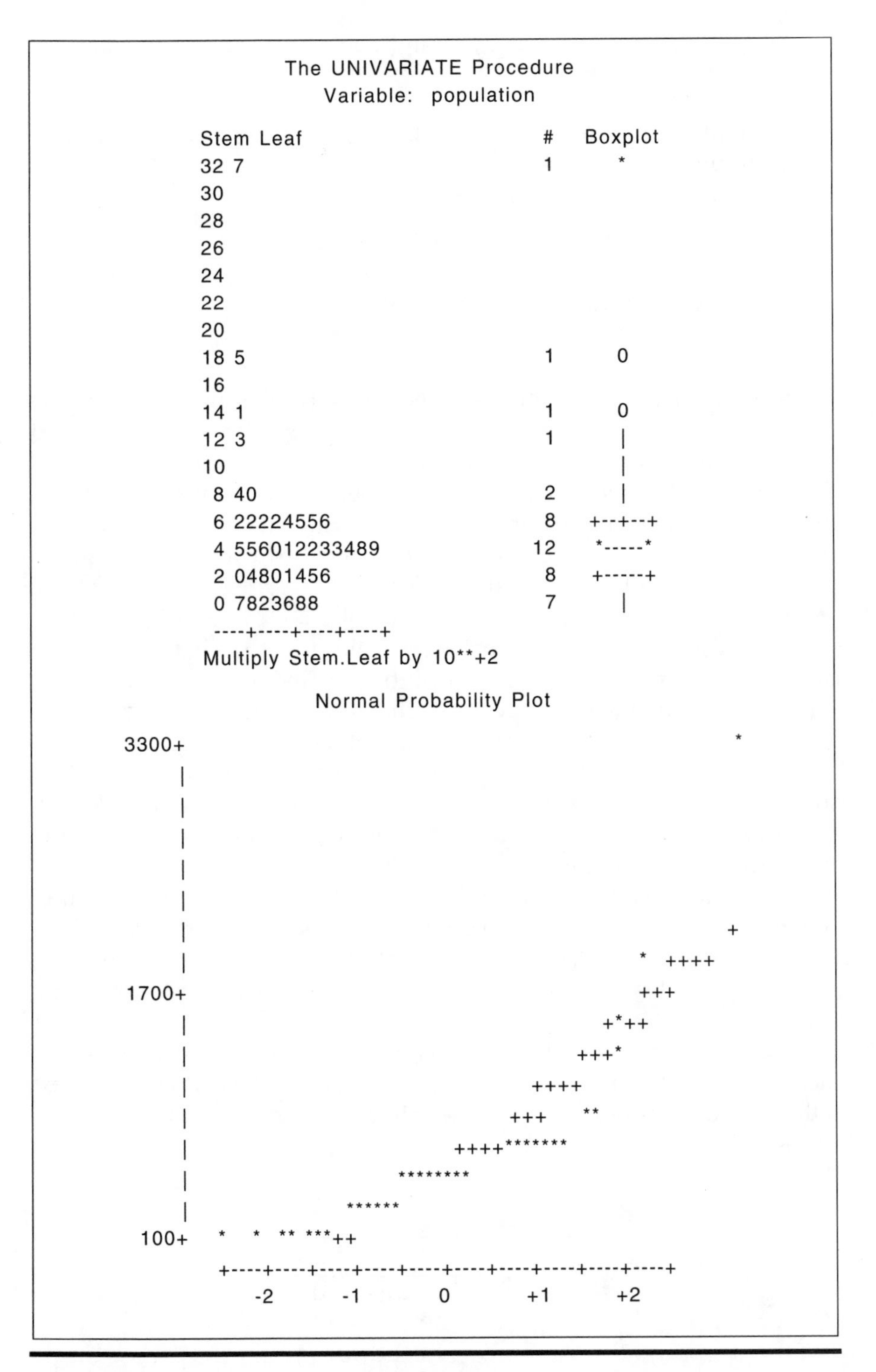

```
                The UNIVARIATE Procedure
                  Variable:  population

        Stem Leaf                       #   Boxplot
          32 7                          1      *
          30
          28
          26
          24
          22
          20
          18 5                          1      0
          16
          14 1                          1      0
          12 3                          1      |
          10                                   |
           8 40                         2      |
           6 22224556                   8   +--+--+
           4 556012233489              12   *-----*
           2 04801456                   8   +-----+
           0 7823688                    7      |
             ----+----+----+----+
        Multiply Stem.Leaf by 10**+2

                  Normal Probability Plot
  3300+                                                     *
      |
      |
      |
      |
      |
      |                                                    +
      |                                              *  ++++
  1700+                                              +++
      |                                            +*++
      |                                          +++*
      |                                       ++++
      |                                    +++    **
      |                                ++++*******
      |                            ********
      |                        ******
   100+   *   *  ** ***++
           +----+----+----+----+----+----+----+----+----+----+
                -2        -1        0        +1        +2
```

Display 14.2

A single linkage cluster analysis and corresponding dendrogram can be obtained as follows:

```
proc cluster data=usair2 method=single simple ccc std out-
tree=single;
   var temperature--rainydays;
   id city;
   copy so2;
proc tree horizontal;
run;
```

The **method=** option in the **proc** statement is self-explanatory. The **simple** option provides information about the distribution of the variables used in the clustering. The **ccc** option includes the cubic clustering criterion in the output, which may be useful for indicating number of groups (Sarle, 1983). The **std** option standardizes the clustering variables to zero mean and unit variance, and the **outtree=** option names the data set that contains the information to be used in the dendrogram.

The **var** statement specifies which variables are to be used to cluster the observations and the **id** statement specifies the variable to be used to label the observations in the printed output and in the dendrogram. Variable(s) mentioned in a **copy** statement are included in the **outtree** data set. Those mentioned in the **var** and **id** statements are included by default.

proc tree produces the dendrogram using the **outtree** data set. The **horizontal** (**hor**) option specifies the orientation, which is vertical by default. The data set to be used by **proc tree** is left implicit and thus will be the most recently created data set (i.e., **single**).

The printed results are shown in Display 14.3 and the dendrogram in Display 14.4. We see that Atlanta and Memphis are joined first to form a two-member group. Then a number of other two-member groups are produced. The first three-member group involves Pittsburgh, Seattle, and Columbus.

First, in Display 14.3 information is provided about the distribution of each variable in the data set. Of particular interest in the clustering context is the bimodality index, which is the following function of skewness and kurtosis:

$$b = \frac{(m_3^2 + 1)}{m_4 + \dfrac{3(n-1)^2}{(n-2)(n-3)}} \tag{14.2}$$

where m_3 is skewness and m_4 is kurtosis. Values of b greater than 0.55 (the value for a uniform population) may indicate bimodal or multimodal marginal distributions. Here, both factories and population have values very close to 0.55, suggesting possible clustering in the data.

The FREQ column of the cluster history simply gives the number of observations in each cluster at each stage of the process. The next two columns, SPRSQ (semipartial *R*-squared) and RSQ (*R*-squared) multiple correlation, are defined as:

$$\text{Semipartial } R^2 = B_{kl}/T \tag{14.3}$$

$$R^2 = 1 - P_g/T \tag{14.4}$$

where $B_{kl} = W_m - W_k - W_l$, with m being the cluster formed from fusing clusters k and l, and W_k is the sum of the distances from each observation in the cluster to the cluster mean; that is:

$$W_k = \sum_{i \in C_k} \|x_i - \bar{x}_k\|^2 \tag{14.5}$$

Finally, $P_g = \Sigma W_j$, where summation is over the number of clusters at the *g*th level of hierarchy.

The single linkage dendrogram in Display 14.4 displays the "chaining" effect typical of this method of clustering. This phenomenon, although somewhat difficult to define formally, refers to the tendency of the technique to incorporate observations into existing clusters, rather than to initiate new ones.

The CLUSTER Procedure
Single Linkage Cluster Analysis

Variable	Mean	Std Dev	Skewness	Kurtosis	Bimodality
temperature	55.5231	6.9762	0.9101	0.7883	0.4525
factories	395.6	330.9	1.9288	5.2670	0.5541
population	538.5	384.0	1.7536	4.3781	0.5341
windspeed	9.5077	1.3447	0.3096	0.2600	0.3120
rain	37.5908	11.0356	-0.6498	1.0217	0.3328
rainydays	115.7	23.9760	-0.1314	0.3393	0.2832

Eigenvalues of the Correlation Matrix

	Eigenvalue	Difference	Proportion	Cumulative
1	2.09248727	0.45164599	0.3487	0.3487
2	1.64084127	0.36576347	0.2735	0.6222
3	1.27507780	0.48191759	0.2125	0.8347
4	0.79316021	0.67485359	0.1322	0.9669
5	0.11830662	0.03817979	0.0197	0.9866
6	0.08012683		0.0134	1.0000

The data have been standardized to mean 0 and variance 1
Root-Mean-Square Total-Sample Standard Deviation = 1
Mean Distance Between Observations = 3.21916

Cluster History

NCL	----------Clusters	Joined----------	FREQ	SPRSQ	RSQ	ERSQ	CCC	Norm Min Dist	T i e
38	Atlanta	Memphis	2	0.0007	.999	.	.	0.1709	
37	Jacksonville	New Orleans	2	0.0008	.998	.	.	0.1919	
36	Des Moines	Omaha	2	0.0009	.998	.	.	0.2023	
35	Nashville	Richmond	2	0.0009	.997	.	.	0.2041	
34	Pittsburgh	Seattle	2	0.0013	.995	.	.	0.236	
33	Louisville	CL35	3	0.0023	.993	.	.	0.2459	
32	Washington	Baltimore	2	0.0015	.992	.	.	0.2577	
31	Columbus	CL34	3	0.0037	.988	.	.	0.2673	
30	CL32	Indianapolis	3	0.0024	.985	.	.	0.2823	
29	CL33	CL31	6	0.0240	.961	.	.	0.3005	
28	CL38	CL29	8	0.0189	.943	.	.	0.3191	
27	CL30	St. Louis	4	0.0044	.938	.	.	0.322	
26	CL27	Kansas City	5	0.0040	.934	.	.	0.3348	
25	CL26	CL28	13	0.0258	.908	.	.	0.3638	
24	Little Rock	CL25	14	0.0178	.891	.	.	0.3651	
23	Minneapolis	Milwaukee	2	0.0032	.887	.	.	0.3775	
22	Hartford	Providence	2	0.0033	.884	.	.	0.3791	
21	CL24	Cincinnati	15	0.0104	.874	.	.	0.3837	
20	CL21	CL36	17	0.0459	.828	.	.	0.3874	
19	CL37	Miami	3	0.0050	.823	.	.	0.4093	
18	CL20	CL22	19	0.0152	.808	.	.	0.4178	
17	Denver	Salt Lake City	2	0.0040	.804	.	.	0.4191	
16	CL18	CL19	22	0.0906	.713	.	.	0.421	
15	CL16	Wilmington	23	0.0077	.705	.	.	0.4257	
14	San Francisco	CL17	3	0.0083	.697	.	.	0.4297	
13	CL15	Albany	24	0.0184	.679	.	.	0.4438	
12	CL13	Norfolk	25	0.0084	.670	.	.	0.4786	

NCL	Clusters Joined		FREQ	SPRSQ	RSQ	ERSQ	CCC	Norm Min Dist	Tie
11	CL12	Wichita	26	0.0457	.625	.	.	0.523	
10	CL14	Albuquerque	4	0.0097	.615	.	.	0.5328	
9	CL23	Cleveland	3	0.0100	.605	.	.	0.5329	
8	CL11	Charleston	27	0.0314	.574	.	.	0.5662	
7	Dallas	Houston	2	0.0078	.566	.731	-6.1	0.5861	
6	CL8	CL9	30	0.1032	.463	.692	-7.6	0.6433	
5	CL6	Buffalo	31	0.0433	.419	.644	-7.3	0.6655	
4	CL5	CL10	35	0.1533	.266	.580	-8.2	0.6869	
3	CL4	CL7	37	0.0774	.189	.471	-6.6	0.6967	

The CLUSTER Procedure
Single Linkage Cluster Analysis

Cluster History

NCL	----------Clusters Joined----------		FREQ	SPRSQ	RSQ	ERSQ	CCC	Norm Min Dist	Tie
2	CL3	Detroit	38	0.0584	.130	.296	-4.0	0.7372	
1	CL2	Philadelphia	39	0.1302	.000	.000	0.00	0.7914	

Display 14.3

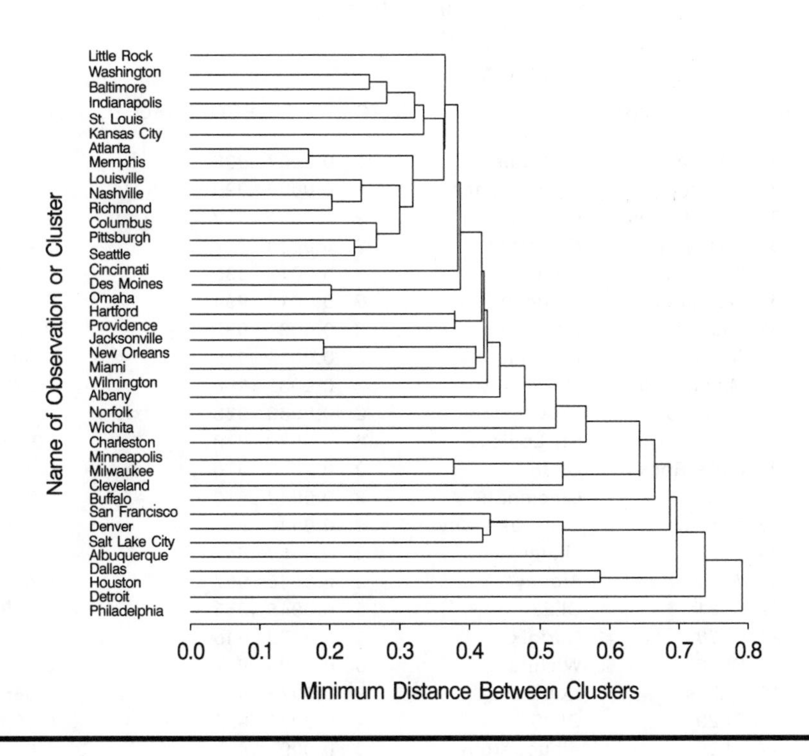

Display 14.4

Resubmitting the SAS code with **method=complete, outree=complete,** and omitting the **simple** option yields the printed results in Display 14.5 and the dendrogram in Display 14.6. Then, substituting **average** for **complete** and resubmitting gives the results shown in Display 14.7 with the corresponding dendrogram in Display 14.8.

```
                    The CLUSTER Procedure
                Complete Linkage Cluster Analysis

              Eigenvalues of the Correlation Matrix

          Eigenvalue    Difference    Proportion    Cumulative

     1    2.09248727    0.45164599        0.3487        0.3487
     2    1.64084127    0.36576347        0.2735        0.6222
     3    1.27507780    0.48191759        0.2125        0.8347
     4    0.79316021    0.67485359        0.1322        0.9669
     5    0.11830662    0.03817979        0.0197        0.9866
     6    0.08012683                      0.0134        1.0000

     The data have been standardized to mean 0 and variance 1
     Root-Mean-Square Total-Sample Standard Deviation = 1
     Mean Distance Between Observations               = 3.21916

                           Cluster History

                                                                Norm T
                                                                 Max i
NCL ----------Clusters Joined---------- FREQ SPRSQ RSQ ERSQ CCC Dist  e

 38 Atlanta          Memphis             2 0.0007 .999 .    .   0.1709
 37 Jacksonville     New Orleans         2 0.0008 .998 .    .   0.1919
 36 Des Moines       Omaha               2 0.0009 .998 .    .   0.2023
 35 Nashville        Richmond            2 0.0009 .997 .    .   0.2041
 34 Pittsburgh       Seattle             2 0.0013 .995 .    .    0.236
 33 Washington       Baltimore           2 0.0015 .994 .    .   0.2577
 32 Louisville       Columbus            2 0.0021 .992 .    .   0.3005
 31 CL33             Indianapolis        3 0.0024 .989 .    .   0.3391
 30 Minneapolis      Milwaukee           2 0.0032 .986 .    .   0.3775
 29 Hartford         Providence          2 0.0033 .983 .    .   0.3791
 28 Kansas City      St. Louis           2 0.0039 .979 .    .    0.412
 27 Little Rock      CL35                3 0.0043 .975 .    .   0.4132
 26 CL32             Cincinnati          3 0.0042 .970 .    .   0.4186
 25 Denver           Salt Lake City      2 0.0040 .967 .    .   0.4191
 24 CL37             Miami               3 0.0050 .962 .    .   0.4217
 23 Wilmington       Albany              2 0.0045 .957 .    .   0.4438
 22 CL31             CL28                5 0.0045 .953 .    .   0.4882
 21 CL38             Norfolk             3 0.0073 .945 .    .   0.5171
 20 CL36             Wichita             3 0.0086 .937 .    .   0.5593
 19 Dallas           Houston             2 0.0078 .929 .    .   0.5861
 18 CL29             CL23                4 0.0077 .921 .    .   0.5936
 17 CL25             Albuquerque         3 0.0090 .912 .    .   0.6291
```

16	CL30	Cleveland	3	0.0100	.902	.	.	0.6667
15	San Francisco	CL17	4	0.0089	.893	.	.	0.6696
14	CL26	CL34	5	0.0130	.880	.	.	0.6935
13	CL27	CL21	6	0.0132	.867	.	.	0.7053
12	CL16	Buffalo	4	0.0142	.853	.	.	0.7463
11	Detroit	Philadelphia	2	0.0142	.839	.	.	0.7914
10	CL18	CL14	9	0.0200	.819	.	.	0.8754
9	CL13	CL22	11	0.0354	.783	.	.	0.938
8	CL10	Charleston	10	0.0198	.763	.	.	1.0649
7	CL15	CL20	7	0.0562	.707	.731	-1.1	1.134
6	CL9	CL8	21	0.0537	.653	.692	-1.6	1.2268
5	CL24	CL19	5	0.0574	.596	.644	-1.9	1.2532
4	CL6	CL12	25	0.1199	.476	.580	-3.2	1.5542
3	CL4	CL5	30	0.1296	.347	.471	-3.2	1.8471
2	CL3	CL7	37	0.1722	.174	.296	-3.0	1.9209
1	CL2	CL11	39	0.1744	.000	.000	0.00	2.3861

Display 14.5

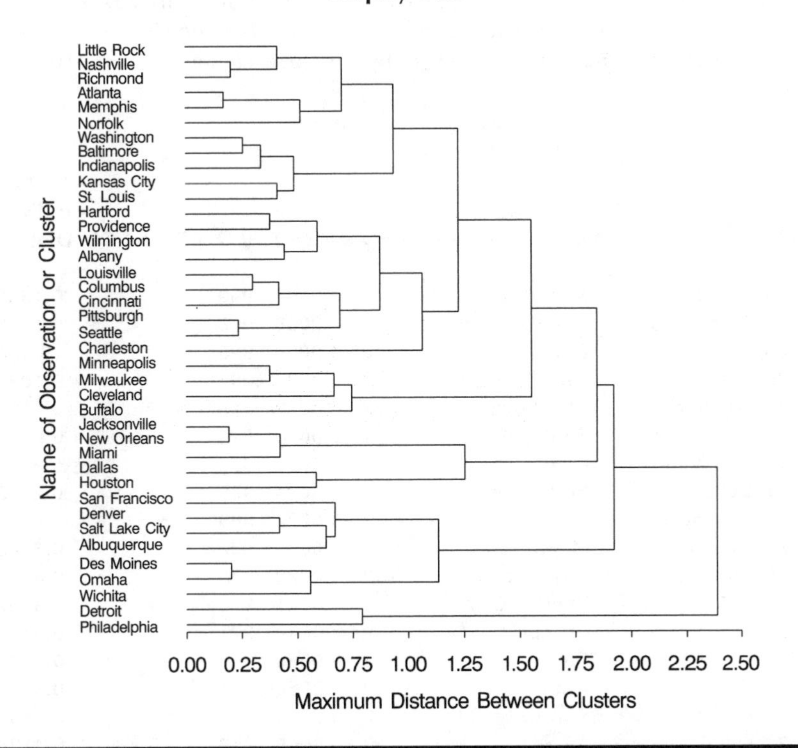

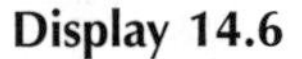
Display 14.6

We see in Display 14.5 that complete linkage clustering begins with the same initial "fusions" of pairs of cities as single linkage, but eventually begins to join different sets of observations. The corresponding dendrogram in Display 14.6 shows a little more structure, although the number of groups is difficult to assess both from the dendrogram and using the CCC criterion.

```
                    The CLUSTER Procedure
               Average Linkage Cluster Analysis

              Eigenvalues of the Correlation Matrix

        Eigenvalue    Difference    Proportion    Cumulative

   1    2.09248727    0.45164599        0.3487        0.3487
   2    1.64084127    0.36576347        0.2735        0.6222
   3    1.27507780    0.48191759        0.2125        0.8347
   4    0.79316021    0.67485359        0.1322        0.9669
   5    0.11830662    0.03817979        0.0197        0.9866
   6    0.08012683                      0.0134        1.0000

   The data have been standardized to mean 0 and variance 1
   Root-Mean-Square Total-Sample Standard Deviation = 1
   Root-Mean-Square Distance Between Observations   = 3.464102

                        Cluster History

                                                                Norm T
                                                                RMS  i
NCL ----------Clusters Joined---------- FREQ SPRSQ  RSQ ERSQ CCC Dist  e

 38 Atlanta         Memphis              2  0.0007 .999 .    .   0.1588
 37 Jacksonville    New Orleans          2  0.0008 .998 .    .   0.1783
 36 Des Moines      Omaha                2  0.0009 .998 .    .    0.188
 35 Nashville       Richmond             2  0.0009 .997 .    .   0.1897
 34 Pittsburgh      Seattle              2  0.0013 .995 .    .   0.2193
 33 Washington      Baltimore            2  0.0015 .994 .    .   0.2395
 32 Louisville      CL35                 3  0.0023 .992 .    .   0.2721
 31 CL33            Indianapolis         3  0.0024 .989 .    .   0.2899
 30 Columbus        CL34                 3  0.0037 .985 .    .    0.342
 29 Minneapolis     Milwaukee            2  0.0032 .982 .    .   0.3508
 28 Hartford        Providence           2  0.0033 .979 .    .   0.3523
 27 CL31            Kansas City          4  0.0041 .975 .    .   0.3607
 26 CL27            St. Louis            5  0.0042 .971 .    .   0.3733
 25 CL32            Cincinnati           4  0.0050 .965 .    .   0.3849
 24 CL37            Miami                3  0.0050 .961 .    .   0.3862
 23 Denver          Salt Lake City       2  0.0040 .957 .    .   0.3894
 22 Wilmington      Albany               2  0.0045 .952 .    .   0.4124
 21 CL38            Norfolk              3  0.0073 .945 .    .    0.463
 20 CL28            CL22                 4  0.0077 .937 .    .   0.4682
 19 CL36            Wichita              3  0.0086 .929 .    .   0.5032
 18 Little Rock     CL21                 4  0.0082 .920 .    .   0.5075
```

```
17 San Francisco  CL23            3  0.0083 .912 .        .     0.5228
16 CL20           CL30            7  0.0166 .896 .        .     0.5368
15 Dallas         Houston         2  0.0078 .888 .        .     0.5446
14 CL18           CL25            8  0.0200 .868 .        .     0.5529
13 CL29           Cleveland       3  0.0100 .858 .        .     0.5608
12 CL17           Albuquerque     4  0.0097 .848 .        .     0.5675
11 CL14           CL26           13  0.0347 .813 .        .     0.6055
10 CL11           CL16           20  0.0476 .766 .        .     0.6578
 9 CL13           Buffalo         4  0.0142 .752 .        .     0.6666
 8 Detroit        Philadelphia    2  0.0142 .737 .        .     0.7355
 7 CL10           Charleston     21  0.0277 .710 .731   -.97    0.8482
 6 CL12           CL19            7  0.0562 .653 .692   -1.6    0.8873
 5 CL7            CL24           24  0.0848 .569 .644   -2.9    0.9135
 4 CL5            CL6            31  0.1810 .388 .580   -5.5    1.0514
 3 CL4            CL9            35  0.1359 .252 .471   -5.3    1.0973
 2 CL3            CL15           37  0.0774 .174 .296   -3.0    1.1161
 1 CL2            CL8            39  0.1744 .000 .000   0.00    1.5159
```

Display 14.7

The average linkage results in Display 14.7 are more similar to those of complete linkage than single linkage; and again, the dendrogram (Display 14.8) suggests more evidence of structure, without making the optimal number of groups obvious.

It is often useful to display the solutions given by a clustering technique by plotting the data in the space of the first two or three principal components and labeling the points by the cluster to which they have been assigned. The number of groups that we should use for these data is not clear from the previous analyses; but to illustrate, we will show the four-group solution obtained from complete linkage.

```
proc tree data=complete out=clusters n=4 noprint;
  copy city so2 temperature--rainydays;
run;
```

As well as producing a dendrogram, proc tree can also be used to create a data set containing a variable, cluster, that indicates to which of a specified number of clusters each observation belongs. The number of clusters is specified with the n= option. The copy statement transfers the named variables to this data set.

The mean vectors of the four groups are also useful in interpretation. These are obtained as follows and the output shown in Display 14.9.

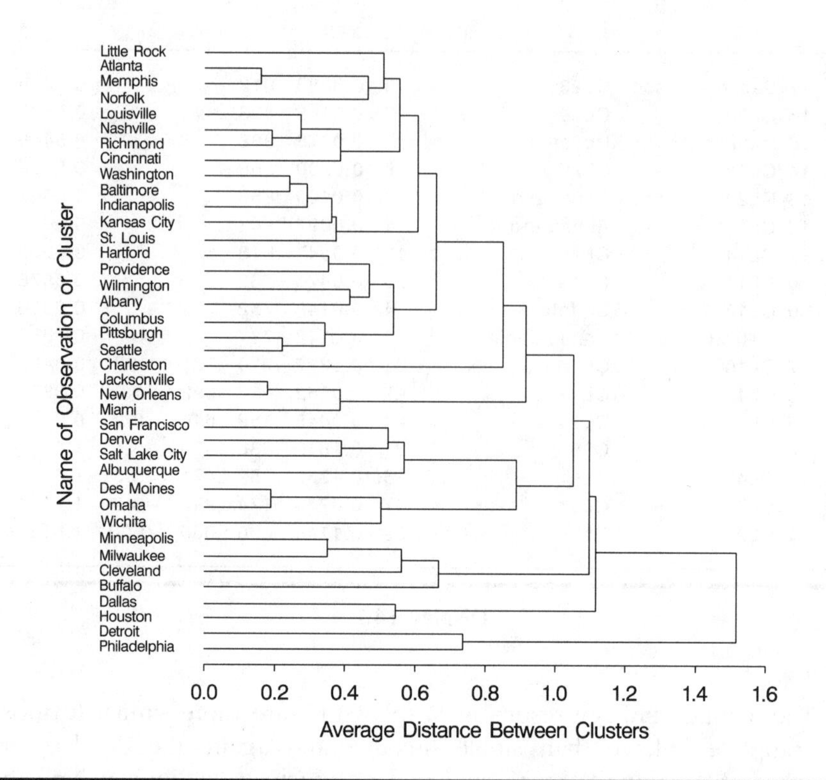

Display 14.8

```
proc sort data=clusters;
   by cluster;

proc means data=clusters;
   var temperature--rainydays;
   by cluster;
run;
```

-- CLUSTER=1 --

The MEANS Procedure

Variable	N	Mean	Std Dev	Minimum	Maximum
temperature	25	53.6040000	5.0301160	43.5000000	61.6000000
factories	25	370.6000000	233.7716122	35.0000000	1007.00
population	25	470.4400000	243.8555037	71.0000000	905.0000000
windspeed	25	9.3240000	1.3690873	6.5000000	12.4000000
rain	25	39.6728000	5.6775101	25.9400000	49.1000000
rainydays	25	125.8800000	18.7401530	99.0000000	166.0000000

CLUSTER=2

Variable	N	Mean	Std Dev	Minimum	Maximum
temperature	5	69.4600000	3.5317135	66.2000000	75.5000000
factories	5	381.8000000	276.0556103	136.0000000	721.0000000
population	5	660.4000000	378.9364063	335.0000000	1233.00
windspeed	5	9.5800000	1.1798305	8.4000000	10.9000000
rain	5	51.0340000	9.4534084	35.9400000	59.8000000
rainydays	5	107.6000000	18.7962762	78.0000000	128.0000000

CLUSTER=3

Variable	N	Mean	Std Dev	Minimum	Maximum
temperature	7	53.3571429	3.2572264	49.0000000	56.8000000
factories	7	214.2857143	168.3168780	46.0000000	454.0000000
population	7	353.7142857	195.8466358	176.0000000	716.0000000
windspeed	7	10.0142857	1.5879007	8.7000000	12.7000000
rain	7	21.1657143	9.5465436	7.7700000	30.8500000
rainydays	7	83.2857143	16.0801564	58.0000000	103.0000000

CLUSTER=4

Variable	N	Mean	Std Dev	Minimum	Maximum
temperature	2	52.2500000	3.3234019	49.9000000	54.6000000
factories	2	1378.00	444.0630586	1064.00	1692.00
population	2	1731.50	309.0056634	1513.00	1950.00
windspeed	2	9.8500000	0.3535534	9.6000000	10.1000000
rain	2	35.4450000	6.3427478	30.9600000	39.9300000
rainydays	2	122.0000000	9.8994949	115.0000000	129.0000000

Display 14.9

A plot of the first two principal components showing cluster membership can be produced as follows. The result is shown in Display 14.10.

```
proc princomp data=clusters n=2 out=pcout noprint;
  var temperature--rainydays;
```

```
proc gplot data=pcout;
   symbol1 v='1';
   symbol2 v='2';
   symbol3 v='3';
   symbol4 v='4';
   plot prin1*prin2=cluster;
run;
```

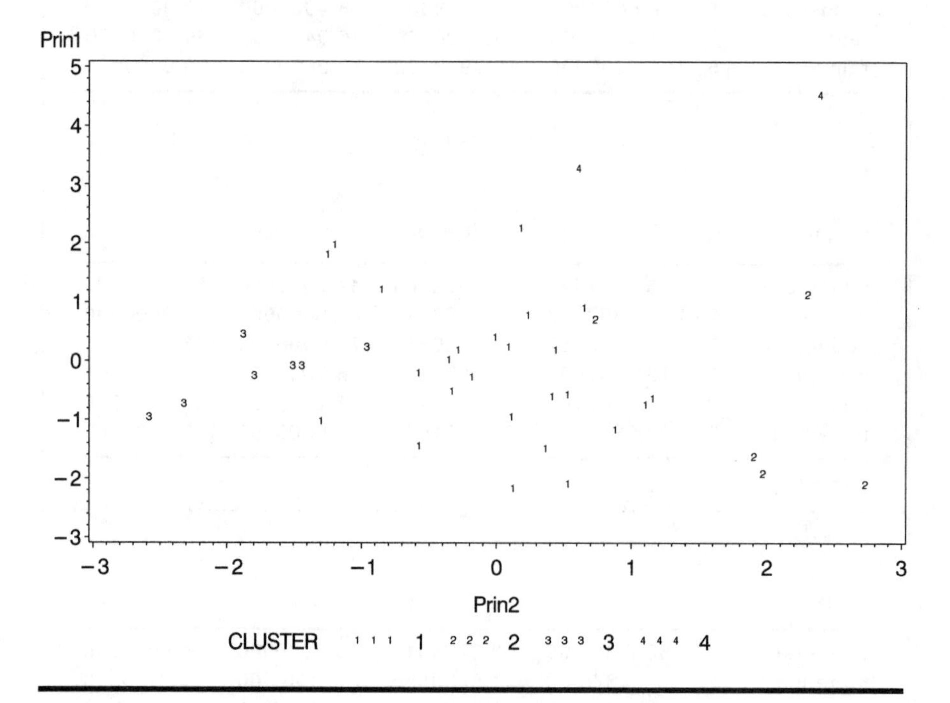

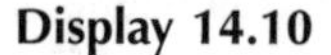

Display 14.10

We see that this solution contains three clusters containing only a few observations each, and is perhaps not ideal. Nevertheless, we continue to use it and look at differences between the derived clusters in terms of differences in air pollution as measured by their average SO_2 values. Box plots of SO_2 values for each of the four clusters can be found as follows:

```
proc boxplot data=clusters;
   plot so2*cluster;
run;
```

The plot is shown in Display 14.11.

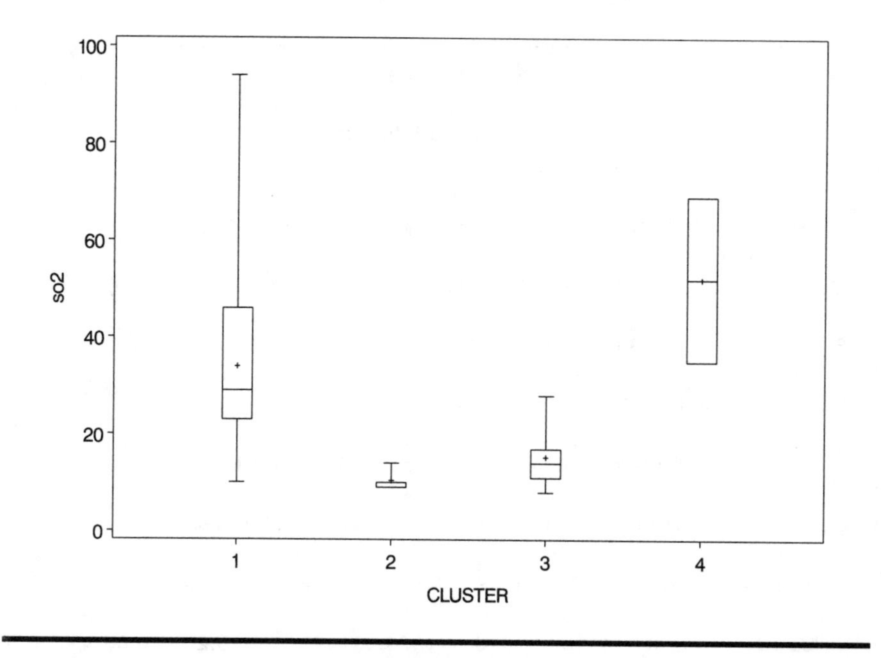

Display 14.11

More formally, we might test for a cluster difference in SO_2 values using a one-way analysis of variance. Here we shall use proc glm for the analysis. The output is shown in Display 14.12.

```
proc glm data=clusters;
   class cluster;
   model so2=cluster;
run;
```

```
                    The GLM Procedure

                 Class Level Information


              Class         Levels    Values

              CLUSTER          4      1 2 3 4

              Number of observations    39
```

The GLM Procedure

Dependent Variable: so2

Source	DF	Sum of Squares	Mean Square	F Value	Pr > F
Model	3	4710.15502	1570.05167	5.26	0.0042
Error	35	10441.58857	298.33110		
Corrected Total	38	15151.74359			

R-Square	Coeff Var	Root MSE	so2 Mean
0.310866	60.57718	17.27226	28.51282

Source	DF	Type I SS	Mean Square	F Value	Pr > F
CLUSTER	3	4710.155018	1570.051673	5.26	0.0042

Source	DF	Type III SS	Mean Square	F Value	Pr > F
CLUSTER	3	4710.155018	1570.051673	5.26	0.0042

Display 14.12

The analysis suggests that there is a significant difference in SO_2 levels in the four clusters.

This probable difference might be investigated in more detail using a suitable multiple comparison procedure (see Exercise 14.4).

Exercises

14.1 Explore some of the other clustering procedures available in SAS; for example, proc modeclus, on the air pollution data.

14.2 Repeat the cluster analysis described in this chapter, but now including the data on Chicago and Phoenix.

14.3 In the cluster analysis described in the text, the data were standardized prior to clustering to unit standard deviation for each variable. Repeat the analysis when the data are standardized by the range instead.

14.4 Use a multiple comparison procedure to investigate which particular clusters in the complete linkage solution differ in air pollution level.

Chapter 15

Discriminant Function Analysis: Classifying Tibetan Skulls

15.1 Description of Data

In the 1920s, Colonel Waddell collected 32 skulls in the southwestern and eastern districts of Tibet. The collection comprised skulls of two types:

- Type A: 17 skulls from graves in Sikkim and neighbouring areas of Tibet
- Type B: 15 skulls picked up on a battlefield in the Lhausa district and believed to be those of native soldiers from the eastern province of Kharis

It was postulated that Tibetans from Kharis might be survivors of a particular fundamental human type, unrelated to the Mongolian and Indian types that surrounded them.

A number of measurements were made on each skull and Display 15.1 shows five of these for each of the 32 skulls collected by Colonel Waddell. (The data are given in Table 144 of *SDS*.) Of interest here is whether the two types of skull can be accurately classified from the five measurements recorded, and which of these five measurements are most informative in this classification exercise.

Type A Skulls

X1	X2	X3	X4	X5
190.5	152.5	145.0	73.5	136.5
172.5	132.0	125.5	63.0	121.0
167.0	130.0	125.5	69.5	119.5
169.5	150.5	133.5	64.5	128.0
175.0	138.5	126.0	77.5	135.5
177.5	142.5	142.5	71.5	131.0
179.5	142.5	127.5	70.5	134.5
179.5	138.0	133.5	73.5	132.5
173.5	135.5	130.5	70.0	133.5
162.5	139.0	131.0	62.0	126.0
178.5	135.0	136.0	71.0	124.0
171.5	148.5	132.5	65.0	146.5
180.5	139.0	132.0	74.5	134.5
183.0	149.0	121.5	76.5	142.0
169.5	130.0	131.0	68.0	119.0
172.0	140.0	136.0	70.5	133.5
170.0	126.5	134.5	66.0	118.5

Type B Skulls

X1	X2	X3	X4	X5
182.5	136.0	138.5	76.0	134.0
179.5	135.0	128.5	74.0	132.0
191.0	140.5	140.5	72.5	131.5
184.5	141.5	134.5	76.5	141.5
181.0	142.0	132.5	79.0	136.5
173.5	136.5	126.0	71.5	136.5
188.5	130.0	143.0	79.5	136.0
175.0	153.0	130.0	76.5	142.0
196.0	142.5	123.5	76.0	134.0
200.0	139.5	143.5	82.5	146.0
185.0	134.5	140.0	81.5	137.0

Type B Skulls (Continued)

X1	X2	X3	X4	X5
174.5	143.5	132.5	74.0	136.5
195.5	144.0	138.5	78.5	144.0
197.0	131.5	135.0	80.5	139.0
182.5	131.0	135.0	68.5	136.0

Note: X1 = greatest length of skull; X2 = greatest horizontal breadth of skull; X3 = height of skull; X4 = upper face height; and X5 = face breadth, between outermost points of cheek bones.

Display 15.1

15.2 Discriminant Function Analysis

Discriminant analysis is concerned with deriving helpful rules for allocating observations to one or another of a set of *a priori* defined classes in some optimal way, using the information provided by a series of measurements made of each sample member. The technique is used in situations in which the investigator has one set of observations, the *training sample*, for which group membership is known with certainty *a priori*, and a second set, the *test sample*, consisting of the observations for which group membership is unknown and which we require to allocate to one of the known groups with as few misclassifications as possible.

An initial question that might be asked is: since the members of the training sample can be classified with certainty, why not apply the procedure used in their classification to the test sample? Reasons are not difficult to find. In medicine, for example, it might be possible to diagnose a particular condition with certainty only as a result of a post-mortem examination. Clearly, for patients still alive and in need of treatment, a different diagnostic procedure would be useful!

Several methods for discriminant analysis are available, but here we concentrate on the one proposed by Fisher (1936) as a method for classifying an observation into one of two possible groups using measurements $x_1, x_2, \cdots, x_p$. Fisher's approach to the problem was to seek a linear function z of the variables:

$$z = a_1x_1 + a_2x_2 + \cdots + a_px_p \quad (15.1)$$

such that the ratio of the between-groups variance of z to its within-group variance is maximized. This implies that the coefficients $\boldsymbol{a}' = [a_1, \cdots, a_p]$ have to be chosen so that V, given by:

$$V = \frac{\boldsymbol{a}'\boldsymbol{Ba}}{\boldsymbol{a}'\boldsymbol{Sa}} \quad (15.2)$$

is maximized. In Eq. (15.2), $\boldsymbol{S}$ is the pooled within-groups covariance matrix; that is

$$\boldsymbol{S} = \frac{(n_1 - 1)\boldsymbol{S}_2 + (n_2 - 1)\boldsymbol{S}_2}{n_1 + n_2 - 2} \quad (15.3)$$

where $\boldsymbol{S}_1$ and $\boldsymbol{S}_2$ are the covariance matrices of the two groups, and n_1 and n_2 the group sample sizes. The matrix $\boldsymbol{B}$ in Eq. (15.2) is the covariance matrix of the group means.

The vector $\boldsymbol{a}$ that maximizes V is given by the solution of the equation:

$$(\boldsymbol{B} - \lambda\boldsymbol{S})\ \boldsymbol{a} = 0 \quad (15.4)$$

In the two-group situation, the single solution can be shown to be:

$$\boldsymbol{a} = \boldsymbol{S}^{-1}(\bar{\boldsymbol{x}}_1 - \bar{\boldsymbol{x}}_2) \quad (15.5)$$

where $\bar{\boldsymbol{x}}_1$ and $\bar{\boldsymbol{x}}_2$ are the mean vectors of the measurements for the observations in each group.

The assumptions under which Fisher's method is optimal are

- The data in both groups have a multivariate normal distribution.
- The covariance matrices of each group are the same.

If the covariance matrices are not the same, but the data are multivariate normal, a quadratic discriminant function may be required. If the data are not multivariate normal, an alternative such as *logistic discrimination* (Everitt and Dunn [2001]) may be more useful, although Fisher's method is known to be relatively robust against departures from normality (Hand [1981]).

Assuming $\bar{z}_1 > \bar{z}_2$, where $\bar{z}_1$ and $\bar{z}_2$ are the discriminant function score means in each group, the classification rule for an observation with discriminant score z_i is:

Assign to group 1 if $z_i - z_c < 0$,
Assign to group 2 if $z_i - z_c \geq 0$,

where

$$z_c = \frac{\bar{z}_1 + \bar{z}_2}{z} \tag{15.6}$$

(This rule assumes that the prior probabilities of belonging to each group are the same.) Subsets of variables most useful for discrimination can be identified using procedures similar to the stepwise methods described in Chapter 4.

A question of some importance about a discriminant function is: how well does it perform? One possible method of evaluating performance would be to apply the derived classification rule to the training set data and calculate the misclassification rate; this is known as the *resubstitution estimate*. However, estimating misclassifications rates in this way, although simple, is known in general to be optimistic (in some cases wildly so). Better estimates of misclassification rates in discriminant analysis can be defined in a variety of ways (see Hand [1997]). One method that is commonly used is the so-called *leaving one out method*, in which the discriminant function is first derived from only n–1 sample members, and then used to classify the observation not included. The procedure is repeated n times, each time omitting a different observation.

15.3 Analysis Using SAS

The data from Display 15.1 can be read in as follows:

```
data skulls;
  infile 'n:\handbook2\datasets\tibetan.dat' expandtabs;
  input length width height faceheight facewidth;
  if _n_ < 18 then type='A';
  else type='B';
run;
```

A parametric discriminant analysis can be specified as follows:

```
proc discrim data=skulls pool=test simple manova wcov cross
  validate;
  class type;
  var length--facewidth;
run;
```

The option **pool=test** provides a test of the equality of the within-group covariance matrices. If the test is significant beyond a level specified by **slpool**, then a quadratic rather than a linear discriminant function is derived. The default value of **slpool** is 0.1,

The **manova** option provides a test of the equality of the mean vectors of the two groups. Clearly, if there is no difference, a discriminant analysis is mostly a waste of time.

The **simple** option provides useful summary statistics, both overall and within groups; **wcov** gives the within-group covariance matrices, the **cross-validate** option is discussed later in the chapter; the **class** statement names the variable that defines the groups; and the **var** statement names the variables to be used to form the discriminant function.

The output is shown in Display 15.2. The results for the test of the equality of the within-group covariance matrices are shown in Display 15.2. The chi-squared test of the equality of the two covariance matrices is not significant at the 0.1 level and thus a linear discriminant function will be derived. The results of the multivariate analysis of variance are also shown in Display 15.2. Because there are only two groups here, all four test criteria lead to the same *F*-value, which is significant well beyond the 5% level.

The results defining the discriminant function are given in Display 15.2. The two sets of coefficients given need to be subtracted to give the discriminant function in the form described in the previous chapter section. This leads to:

$$\boldsymbol{a}' = [-0.0893,\ 0.1158,\ 0.0052,\ -0.1772,\ -0.1774] \qquad (15.7)$$

The group means on the discriminant function are $\bar{z}_1 = -28.713$, $\bar{z}_2 = -32.214$, leading to a value of $\bar{z}_c = -30.463$.

Thus, for example, a skull having a vector of measurements $x' = [185, 142, 130, 72, 133]$ has a discriminant score of –30.07, and $z_1 - z_c$ in this case is therefore 0.39 and the skull should be assigned to group 1.

```
                     The DISCRIM Procedure

        Observations    32    DF Total                31
        Variables        5    DF Within Classes       30
        Classes          2    DF Between Classes       1

                     Class Level Information

          Variable                                             Prior
    type  Name        Frequency     Weight   Proportion  Probability

    A     A                  17    17.0000     0.531250     0.500000
    B     B                  15    15.0000     0.468750     0.500000
```

The DISCRIM Procedure
Within-Class Covariance Matrices

type = A, DF = 16

Variable	length	width	height	faceheight	facewidth
length	45.52941176	25.22242647	12.39062500	22.15441176	27.97242647
width	25.22242647	57.80514706	11.87500000	7.51930147	48.05514706
height	12.39062500	11.87500000	36.09375000	-0.31250000	1.40625000
faceheight	22.15441176	7.51930147	-0.31250000	20.93566176	16.76930147
facewidth	27.97242647	48.05514706	1.40625000	16.76930147	66.21139706

---

type = B, DF = 14

Variable	length	width	height	faceheight	facewidth
length	74.42380952	-9.52261905	22.73690476	17.79404762	11.12500000
width	-9.52261905	37.35238095	-11.26309524	0.70476190	9.46428571
height	22.73690476	-11.26309524	36.31666667	10.72380952	7.19642857
faceheight	17.79404762	0.70476190	10.72380952	15.30238095	8.66071429
facewidth	11.12500000	9.46428571	7.19642857	8.66071429	17.96428571

The DISCRIM Procedure
Simple Statistics

Total-Sample

Variable	N	Sum	Mean	Variance	Standard Deviation
length	32	5758	179.93750	87.70565	9.3651
width	32	4450	139.06250	46.80242	6.8412
height	32	4266	133.29688	36.99773	6.0826
faceheight	32	2334	72.93750	29.06048	5.3908
facewidth	32	4279	133.70313	55.41709	7.4443

---

type = A

Variable	N	Sum	Mean	Variance	Standard Deviation
length	17	2972	174.82353	45.52941	6.7475
width	17	2369	139.35294	57.80515	7.6030
height	17	2244	132.00000	36.09375	6.0078
faceheight	17	1187	69.82353	20.93566	4.5756
facewidth	17	2216	130.35294	66.21140	8.1370

---

type = B

Variable	N	Sum	Mean	Variance	Standard Deviation
length	15	2786	185.73333	74.42381	8.6269
width	15	2081	138.73333	37.35238	6.1117
height	15	2022	134.76667	36.31667	6.0263
faceheight	15	1147	76.46667	15.30238	3.9118
facewidth	15	2063	137.50000	17.96429	4.2384

Within Covariance Matrix Information

type	Covariance Matrix Rank	Natural Log of the Determinant of the Covariance Matrix
A	5	16.16370
B	5	15.77333
Pooled	5	16.72724

The DISCRIM Procedure
Test of Homogeneity of Within Covariance Matrices

Notation: K = Number of Groups

P = Number of Variables

N = Total Number of Observations - Number of Groups

N(i) = Number of Observations in the i'th Group - 1

```
                                          N(i)/2
                  __
                 | |  |within SS Matrix (i)|
      V      =   -----------------------------------
                                          N/2
                        |Pooled SS Matrix|

                                                  2
                       |¯      1      1   ¯|  2P + 3P - 1
      RHO    =  1.0 - |  SUM ----- - ---   |  ----------------
                       |_      N(i)   N   _|  6 (P+1) (K-1)

      DF     =  .5(K-1)P(P+1)

                                       |¯   PN/2          ¯|
                                       |   N          V    |
Under the null hypothesis: -2 RHO ln   |  ---------------- |
                                       |    __      PN(i)/2|
                                       |_  | | N(i)       _|

is distributed approximately as Chi-Square(DF).

                    Chi-Square    DF    Pr > ChiSq

                     18.370512    15        0.2437

Since the Chi-Square value is not significant at the 0.1 level, a pooled
covariance matrix will be used in the discriminant function.
Reference: Morrison, D.F. (1976) Multivariate Statistical
Methods p252.

                      The DISCRIM Procedure

       Pairwise Generalized Squared Distances Between Groups

                   2                    -1
                  D (i|j) = (X̄  - X̄)' COV   (X̄  - X̄)
                              i    j           i    j

              Generalized Squared Distance to type

                    From
                    type            A           B

                    A               0     3.50144
                    B         3.50144           0
```

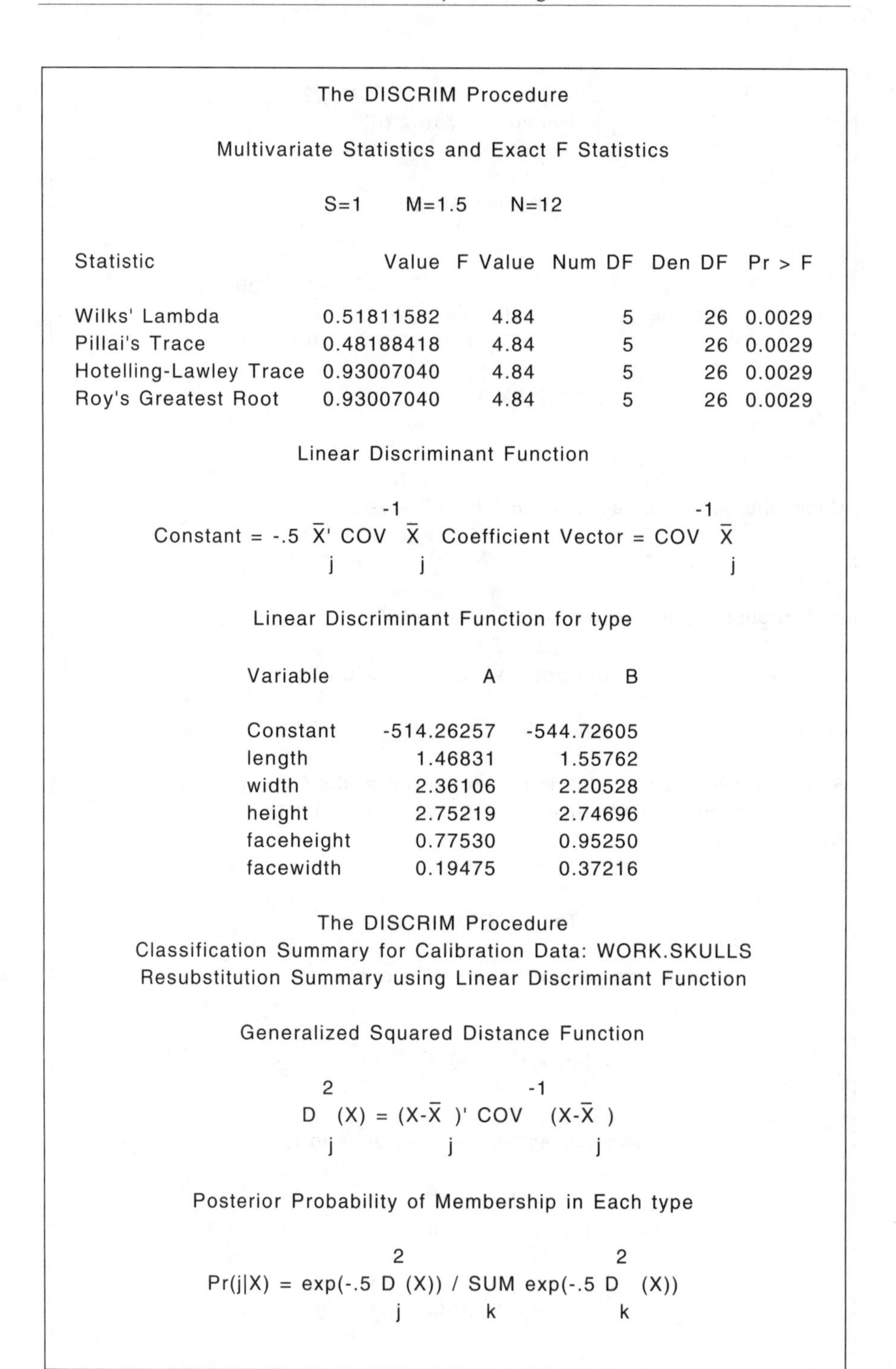

```
                      The DISCRIM Procedure

          Multivariate Statistics and Exact F Statistics

                     S=1      M=1.5      N=12

Statistic                      Value  F Value  Num DF  Den DF  Pr > F

Wilks' Lambda             0.51811582     4.84       5      26  0.0029
Pillai's Trace            0.48188418     4.84       5      26  0.0029
Hotelling-Lawley Trace    0.93007040     4.84       5      26  0.0029
Roy's Greatest Root       0.93007040     4.84       5      26  0.0029

                  Linear Discriminant Function

                        -1                                -1
      Constant = -.5 X' COV  X   Coefficient Vector = COV  X
                      j       j                             j

             Linear Discriminant Function for type

            Variable                  A              B

            Constant         -514.26257     -544.72605
            length              1.46831        1.55762
            width               2.36106        2.20528
            height              2.75219        2.74696
            faceheight          0.77530        0.95250
            facewidth           0.19475        0.37216

                      The DISCRIM Procedure
     Classification Summary for Calibration Data: WORK.SKULLS
     Resubstitution Summary using Linear Discriminant Function

              Generalized Squared Distance Function

                     2                  -1
                    D  (X) = (X-X )' COV   (X-X )
                     j           j              j

          Posterior Probability of Membership in Each type

                            2                  2
          Pr(j|X) = exp(-.5 D (X)) / SUM exp(-.5 D  (X))
                            j      k           k
```

```
Number of Observations and Percent Classified into type

          From
          type          A          B      Total

          A            14          3         17
                    82.35      17.65     100.00

          B             3         12         15
                    20.00      80.00     100.00

          Total        17         15         32
                    53.13      46.88     100.00

          Priors      0.5        0.5


            Error Count Estimates for type


                        A          B      Total

          Rate     0.1765     0.2000     0.1882
          Priors   0.5000     0.5000


                   The DISCRIM Procedure
   Classification Summary for Calibration Data: WORK.SKULLS
   Cross-validation Summary using Linear Discriminant Function


           Generalized Squared Distance Function

                 2                   -1
               D  (X) = (X-X̄   )' COV   (X-X̄    )
                 j           (X)j      (X)    (X)j


       Posterior Probability of Membership in Each type

                            2                    2
         Pr(j|X) = exp(-.5 D (X)) / SUM exp(-.5 D (X))
                            j        k           k
```

Number of Observations and Percent Classified into type

From type	A	B	Total
A	12	5	17
	70.59	29.41	100.00
B	6	9	15
	40.00	60.00	100.00
Total	8	14	32
	56.25	43.75	100.00
Priors	0.5	0.5	

Error Count Estimates for type

	A	B	Total
Rate	0.2941	0.4000	0.3471
Priors	0.5000	0.5000	

Display 15.2

The resubstitution approach to estimating the misclassification rate of the derived allocation rule is seen from Display 15.2 to be 18.82%. But the leaving-out-one (cross-validation) approach increases this to a more realistic 34.71%.

To identify the most important variables for discrimination, proc stepdisc can be used as follows. The output is shown in Display 15.3.

```
proc stepdisc data=skulls sle=.05 sls=.05;
   class type;
   var length--facewidth;
run;
```

The significance levels required for variables to enter and be retained are set with the sle (slentry) and sls (slstay) options, respectively. The default value for both is p=.15. By default, a "stepwise" procedure is used (other options can be specified using a method= statement). Variables are chosen to enter or leave the discriminant function according to one of two criteria:

- The significance level of an *F*-test from an analysis of covariance, where the variables already chosen act as covariates and the variable under consideration is the dependent variable.
- The squared multiple correlation for predicting the variable under consideration from the class variable controlling for the effects of the variables already chosen.

The significance level and the squared partial correlation criteria select variables in the same order, although they may select different numbers of variables. Increasing the sample size tends to increase the number of variables selected when using significance levels, but has little effect on the number selected when using squared partial correlations.

At step 1 in Display 15.3, the variable **faceheight** has the highest R^2 value and is the first variable selected. At Step 2, none of the partial R^2 values of the other variables meet the criterion for inclusion and the process therefore ends. The tolerance shown for each variable is one minus the squared multiple correlation of the variable with the other variables already selected. A variable can only be entered if its tolerance is above a value specified in the **singular** statement. The value set by default is 1.0E–8.

```
                  The STEPDISC Procedure

       The Method for Selecting Variables is STEPWISE


  Observations    32   Variable(s) in the Analysis       5
  Class Levels     2   Variable(s) will be Included      0
                       Significance Level to Enter    0.05
                       Significance Level to Stay     0.05

                  Class Level Information


             Variable
 Type            Name   Frequency     Weight   Proportion

 A                  A          17    17.0000     0.531250
 B                  B          15    15.0000     0.468750
```

The STEPDISC Procedure
Stepwise Selection: Step 1

Statistics for Entry, DF = 1, 30

Variable	R-Square	F Value	Pr > F	Tolerance
length	0.3488	16.07	0.0004	1.0000
width	0.0021	0.06	0.8029	1.0000
height	0.0532	1.69	0.2041	1.0000
faceheight	0.3904	19.21	0.0001	1.0000
facewidth	0.2369	9.32	0.0047	1.0000

Variable faceheight will be entered.

Variable(s) that have been Entered

faceheight

Multivariate Statistics

Statistic	Value	F Value	Num DF	Den DF	Pr > F
Wilks' Lambda	0.609634	19.21	1	30	0.0001
Pillai's Trace	0.390366	9.21	1	30	0.0001
Average Squared Canonical Correlation	0.390366				

The STEPDISC Procedure
Stepwise Selection: Step 2

Statistics for Removal, DF = 1, 30

Variable	R-Square	F Value	Pr > F
faceheight	0.3904	19.21	0.0001

No variables can be removed.

Statistics for Entry, DF = 1, 29

Variable	Partial R-Square	F Value	Pr > F	Tolerance
length	0.0541	1.66	0.2081	0.4304
width	0.0162	0.48	0.4945	0.9927
height	0.0047	0.14	0.7135	0.9177
facewidth	0.0271	0.81	0.3763	0.6190

```
                    No variables can be entered.

                    No further steps are possible.

                       The STEPDISC Procedure

                     Stepwise Selection Summary

                                                                           Averaged
                                                                           Squared
      Number                   Partial      F      Pr >      Wilks'     Pr <     Canonical      Pr >
Step  In Entered     Removed  R-Square   Value        F      Lambda   Lambda   Correlation     ASCC

   1   1 faceheight             0.3904   19.21   0.0001  0.60963388   0.0001    0.39036612   0.0001
```

Display 15.3

Details of the "discriminant function" using only faceheight are found as follows:

```
proc discrim data=skulls crossvalidate;
   class type;
   var faceheight;
run;
```

The output is shown in Display 15.4. Here, the coefficients of faceheight in each class are simply the mean of the class on faceheight divided by the pooled within-group variance of the variable. The resubstitution and leaving one out methods of estimating the misclassification rate give the same value of 24.71%.

```
                       The DISCRIM Procedure

         Observations     32     DF Total                  31
         Variables         1     DF Within Classes         30
         Classes           2     DF Between Classes         1

                      Class Level Information

        Variable                                                 Prior
type    Name        Frequency      Weight    Proportion    Probability

A       A                  17     17.0000      0.531250       0.500000
B       B                  15      5.0000      0.468750       0.500000
```

```
                Pooled Covariance Matrix Information

                            Natural Log of the
                Covariance   Determinant of the
                Matrix Rank   Covariance Matrix

                         1            2.90727

                     The DISCRIM Procedure

       Pairwise Generalized Squared Distances Between Groups

                 2                     -1
                D (i|j) = (X̄  -  X̄)' COV   (X̄  -  X̄)
                            i    j           i    j

                Generalized Squared Distance to type

                    From
                    type          A          B

                    A             0    2.41065
                    B       2.41065          0

                   Linear Discriminant Function

                          -1                               -1
       Constant = -.5  X̄' COV  X̄   Coefficient Vector = COV    X̄
                        j       j                                j

                Linear Discriminant Function for type

               Variable              A             B

               Constant     -133.15615    -159.69891
               faceheight      3.81408       4.17695

                     The DISCRIM Procedure
       Classification Summary for Calibration Data: WORK.SKULLS
       Resubstitution Summary using Linear Discriminant Function

               Generalized Squared Distance Function

                  2                  -1
                 D (X) = (X-X̄ )' COV   (X-X̄ )
                   j         j             j
```

Posterior Probability of Membership in Each type

$$Pr(j|X) = \exp(-.5\, D_j^2(X)) \,/\, \mathrm{SUM}_k \exp(-.5\, D_k^2(X))$$

Number of Observations and Percent Classified into type

From type	A	B	Total
A	12	5	17
	70.59	29.41	100.00
B	3	12	15
	20.00	80.00	100.00
Total	15	17	32
	46.88	53.13	100.00
Priors	0.5	0.5	

Error Count Estimates for type

	A	B	Total
Rate	0.2941	0.2000	0.2471
Priors	0.5000	0.5000	

The DISCRIM Procedure
Classification Summary for Calibration Data: WORK.SKULLS
Cross-validation Summary using Linear Discriminant Function

Generalized Squared Distance Function

$$D_j^2(X) = (X-\overline{X}_{(X)j})' \, COV_{(X)}^{-1} \, (X-\overline{X}_{(X)j})$$

Posterior Probability of Membership in Each type

$$Pr(j|X) = \exp(-.5\, D_j^2(X)) \,/\, \mathrm{SUM}_k \exp(-.5\, D_k^2(X))$$

```
Number of Observations and Percent Classified into type

From
type            A        B      Total

A              12        5         17
            70.59    29.41     100.00

B               3       12         15
            20.00    80.00     100.00

Total          15       17         32
            46.88    53.13     100.00

Priors        0.5      0.5

Error Count Estimates for type

                A          B      Total

Rate       0.2941     0.2000     0.2471
Priors     0.5000     0.5000
```

Display 15.4

Exercises

15.1 Use the posterr options in proc discrim to estimate error rates for the discriminant functions derived for the skull data. Compare these with those given in Displays 15.2 and 15.4.

15.2 Investigate the use of the nonparametric discriminant methods available in proc discrim for the skull data. Compare the results with those for the simple linear discriminant function given in the text.

Chapter 16

Correspondence Analysis: Smoking and Motherhood, Sex and the Single Girl, and European Stereotypes

16.1 Description of Data

Three sets of data are considered in this chapter, all of which arise in the form of two-dimensional contingency tables as met previously in Chapter 3. The three data sets are given in Displays 16.1, 16.2, and 16.3; details are as follows.

- Display 16.1: These data involve the association between a girl's age and her relationship with her boyfriend.
- Display 16.2: These data show the distribution of birth outcomes by age of mother, length of gestation, and whether or not the mother smoked during the prenatal period. We consider the data as a two-dimensional contingency table with four row categories and four column categories.

- Display 16.3: These data were obtained by asking a large number of people in the U.K. which of 13 characteristics they would associate with the nationals of the U.K.'s partner countries in the European Community. Entries in the table give the percentages of respondents agreeing that the nationals of a particular country possess the particular characteristic.

	Age Group				
	Under 16	*16–17*	*17–18*	*18–19*	*19–20*
No boyfriend	21	21	14	13	8
Boyfriend/No sexual intercourse	8	9	6	8	2
Boyfriend/Sexual intercourse	2	3	4	10	10

Display 16.1

	Premature		*Full-Term*	
	Died in 1st year	*Alive at year 1*	*Died in 1st year*	*Alive at year 1*
Young mothers				
Non-smokers	50	315	24	4012
Smokers	9	40	6	459
Old mothers				
Non-smokers	41	147	14	1594
Smokers	4	11	1	124

Display 16.2

	Characteristic												
Country	*1*	*2*	*3*	*4*	*5*	*6*	*7*	*8*	*9*	*10*	*11*	*12*	*13*
France	37	29	21	19	10	10	8	8	6	6	5	2	1
Spain	7	14	8	9	27	7	3	7	3	23	12	1	3
Italy	30	12	19	10	20	7	12	6	5	13	10	1	2
U.K.	9	14	4	6	27	12	2	13	26	16	29	6	25
Ireland	1	7	1	16	30	3	10	9	5	11	22	2	27
Holland	5	4	2	2	15	2	0	13	24	1	28	4	6
Germany	4	48	1	12	3	9	2	11	41	1	38	8	8

Note: Characteristics: (1) stylish; (2) arrogant; (3) sexy; (4) devious; (5) easy-going; (6) greedy; (7) cowardly; (8) boring; (9) efficient; (10) lazy; (11) hard working; (12) clever; (13) courageous.

Display 16.3

16.2 Displaying Contingency Table Data Graphically Using Correspondence Analysis

Correspondence analysis is a technique for displaying the associations among a set of categorical variables in a type of scatterplot or map, thus allowing a visual examination of the structure or pattern of these associations. A correspondence analysis should ideally be seen as an extremely useful supplement to, rather than a replacement for, the more formal inferential procedures generally used with categorical data (see Chapters 3 and 8). The aim when using correspondence analysis is nicely summarized in the following quotation from Greenacre (1992):

> An important aspect of correspondence analysis which distinguishes it from more conventional statistical methods is that it is not a confirmatory technique, trying to prove a hypothesis, but rather an exploratory technique, trying to reveal the data content. One can say that it serves as a window onto the data, allowing researchers easier access to their numerical results and facilitating discussion of the data and possibly generating hypothesis which can be formally tested at a later stage.

Mathematically, correspondence analysis can be regarded as either:

- A method for decomposing the chi-squared statistic for a contingency table into components corresponding to different dimensions of the heterogeneity between its rows and columns, or
- A method for simultaneously assigning a scale to rows and a separate scale to columns so as to maximize the correlation between the resulting pair of variables.

Quintessentially, however, correspondence analysis is a technique for displaying multivariate categorical data graphically, by deriving coordinate values to represent the categories of the variables involved, which can then be plotted to provide a "picture" of the data.

In the case of two categorical variables forming a two-dimensional contingency table, the required coordinates are obtained from the singular value decomposition (Everitt and Dunn [2001]) of a matrix $\boldsymbol{E}$ with elements e_{ij} given by:

$$e_{ij} = \frac{p_{ij} - p_{i+}p_{+j}}{\sqrt{p_{i+}p_{+j}}} \tag{16.1}$$

where $p_{ij} = n_{ij}/n$ with n_{ij} being the number of observations in the ijth cell of the contingency table and n the total number of observations. The total number of observations in row i is represented by n_{i+} and the corresponding value for column j is n_{+j}. Finally $p_{i+} = \frac{n_{i+}}{n}$ and $p_{+j} = \frac{n_{+j}}{n}$. The elements of $\boldsymbol{E}$ can be written in terms of the familiar "observed" (O) and "expected" (E) nomenclature used for contingency tables as:

$$e_{ij} = \frac{1}{\sqrt{n}} \frac{\mathrm{O} - \mathrm{E}}{\sqrt{\mathrm{E}}} \tag{16.2}$$

Written in this way, it is clear that the terms are a form of residual from fitting the independence model to the data.

The singular value decomposition of $\boldsymbol{E}$ consists of finding matrices $\boldsymbol{U}$, $\boldsymbol{V}$, and $\boldsymbol{\Delta}$ (diagonal) such that:

$$\boldsymbol{E} = \boldsymbol{U\Delta V}' \tag{16.3}$$

where $\boldsymbol{U}$ contains the eigenvectors of $\boldsymbol{EE}'$ and $\boldsymbol{V}$ the eigenvectors of $\boldsymbol{E'E}$. The diagonal matrix $\boldsymbol{\Delta}$ contains the ranked singular values δ_k so that δ_k^2 are the eigenvalues (in decreasing) order of either $\boldsymbol{EE}'$ or $\boldsymbol{E'E}$.

The coordinate of the *i*th row category on the *k*th coordinate axis is given by $\delta_k u_{ik}/\sqrt{p_{i+}}$, and the coordinate of the *j*th column category on the same axis is given by $\delta_k v_{jk}/\sqrt{p_{+j}}$, where u_{ik}, $i = 1 \cdots r$ and v_{jk}, $j = 1 \cdots c$ are, respectively, the elements of the *k*th column of $\boldsymbol{U}$ and the *k*th column of $\boldsymbol{V}$.

To represent the table fully requires at most $R = \min(r, c) - 1$ dimensions, where r and c are the number of rows and columns of the table. R is the rank of the matrix $\boldsymbol{E}$. The eigenvalues, δ_k^2, are such that:

$$\text{Trace } (\boldsymbol{EE'}) = \sum_{k=1}^{R} \delta_k^2 = \sum_{i=1}^{r}\sum_{j=1}^{c} e_{ij}^2 = \frac{X^2}{n} \tag{16.4}$$

where X^2 is the usual chi-squared test statistic for independence. In the context of correspondence analysis, X^2/n is known as *inertia*. Correspondence analysis produces a graphical display of the contingency table from the columns of $\boldsymbol{U}$ and $\boldsymbol{V}$, in most cases from the first two columns, $\boldsymbol{u}_1$, $\boldsymbol{u}_2$, $\boldsymbol{v}_1$, $\boldsymbol{v}_2$, of each, since these give the "best" two-dimensional representation. It can be shown that the first two coordinates give the following approximation to the e_{ij}:

$$e_{ij} \approx u_{i1}v_{j1} + u_{i2}v_{j2} \tag{16.5}$$

so that a large positive residual corresponds to u_{ik} and v_{jk} for $k = 1$ or 2, being large and of the same sign. A large negative residual corresponds to u_{ik} and v_{jk}, being large and of opposite sign for each value of k. When u_{ik} and v_{jk} are small and their signs are not consistent for each k, the corresponding residual term will be small. The adequacy of the representation produced by the first two coordinates can be informally assessed by calculating the percentages of the inertia they account for; that is

$$\text{Percentage inertia} = \frac{\delta_1^2 + \delta_2^2}{\sum_{k=1}^{R} \delta_k^2} \tag{16.6}$$

Values of 60% and above usually mean that the two-dimensional solution gives a reasonable account of the structure in the table.

16.3 Analysis Using SAS

16.3.1 Boyfriends

Assuming that the 15 cell counts shown in Display 16.1 are in an ASCII file, tab separated, a suitable data set can be created as follows:

```
data boyfriends;
   infile 'n:\handbook2\datasets\boyfriends.dat' expandtabs;
   input c1-c5;
   if _n_=1 then rowid='NoBoy';
   if _n_=2 then rowid='NoSex';
   if _n_=3 then rowid='Both';
   label c1='under 16' c2='16-17' c3='17-18' c4='18-19'
   c5='19-20';
run;
```

The data are already in the form of a contingency table and can be simply read into a set of variables representing the columns of the table. The label statement is used to assign informative labels to these variables. More informative variable names could also have been used, but labels are more flexible in that they may begin with numbers and include spaces. It is also useful to label the rows, and here the SAS automatic variable _n_ is used to set the values of a character variable rowid.

A correspondence analysis of this table can be performed as follows:

```
proc corresp data=boyfriends out=coor;
   var c1-c5;
   id rowid;
run;
```

The out= option names the data set that will contain the coordinates of the solution. By default, two dimensions are used and the dimens= option is used to specify an alternative.

The var statement specifies the variables that represent the columns of the contingency table, and the id statement specifies a variable to be used to label the rows. The latter is optional, but without it the rows will simply be labelled row1, row2, etc. The output appears in Display 16.4.

```
                    The CORRESP Procedure

            Inertia and Chi-Square Decomposition

Singular  Principal    Chi-               Cumulative
   Value    Inertia  Square  Percent         Percent   19   38   57   76   95
                                                      ----+----+----+----+----+---
 0.37596    0.14135 19.6473    95.36           95.36 ************************
 0.08297    0.00688  0.9569     4.64          100.00 *

   Total    0.14823 20.6042   100.00

Degrees of Freedom = 8

                         Row Coordinates

                              Dim1        Dim2

                  NoBoy    -0.1933     -0.0610
                  NoSex    -0.1924      0.1425
                  Both      0.7322     -0.0002

            Summary Statistics for the Row Points

                        Quality      Mass    Inertia

               NoBoy     1.0000    0.5540     0.1536
               NoSex     1.0000    0.2374     0.0918
               Both      1.0000    0.2086     0.7546

        Partial Contributions to Inertia for the Row Points

                               Dim1       Dim2

                   NoBoy     0.1465     0.2996
                   NoSex     0.0622     0.7004
                   Both      0.7914     0.0000

Indices of the Coordinates that Contribute Most to Inertia for the Row Points

                           Dim1    Dim2    Best

                 NoBoy        2       2       2
                 NoSex        0       2       2
                 Both         1       0       1
```

Squared Cosines for the Row Points

	Dim1	Dim2
NoBoy	0.9094	0.0906
NoSex	0.6456	0.3544
Both	1.0000	0.0000

Column Coordinates

	Dim1	Dim2
under 16	-0.3547	-0.0550
16-17	-0.2897	0.0003
17-18	-0.1033	0.0001
18-19	0.2806	0.1342
19-20	0.7169	-0.1234

The CORRESP Procedure

Summary Statistics for the Column Points

	Quality	Mass	Inertia
under 16	1.0000	0.2230	0.1939
16-17	1.0000	0.2374	0.1344
17-18	1.0000	0.1727	0.0124
18-19	1.0000	0.2230	0.1455
19-20	1.0000	0.1439	0.5137

Partial Contributions to Inertia for the Column Points

	Dim1	Dim2
under 16	0.1985	0.0981
16-17	0.1410	0.0000
17-18	0.0130	0.0000
18-19	0.1242	0.5837
19-20	0.5232	0.3183

Indices of the Coordinates that Contribute Most to Inertia for the Column Points

	Dim1	Dim2	Best
under 16	1	0	1
16-17	1	0	1
17-18	0	0	1
18-19	0	2	2
19-20	1	1	1

Squared Cosines for the Column Points

	Dim1	Dim2
under 16	0.9765	0.0235
16-17	1.0000	0.0000
17-18	1.0000	0.0000
18-19	0.8137	0.1863
19-20	0.9712	0.0288

Display 16.4

To produce a plot of the results, a built-in SAS macro, plotit, can be used:

```
%plotit(data=coor,datatype=corresp,color=black,colors=black);
```

The resulting plot is shown in Display 16.5. Displaying the categories of a contingency table in a scatterplot in this way involves the concept of *distance* between the percentage profiles of row or column categories. The distance measure used in a correspondence analysis is known as the *chi-squared distance.* The calculation of this distance can be illustrated using the proportions of girls in age groups 1 and 2 for each relationship type in Display 16.1.

Chi-squared distance =

$$\sqrt{\frac{(0.68-0.64)^2}{0.55}+\frac{(0.26-0.27)^2}{0.24}+\frac{(0.06-0.09)^2}{0.21}} = 0.09 \qquad (16.7)$$

This is similar to ordinary "straight line" or Pythagorean distance, but differs by dividing each term by the corresponding average proportion.

In this way, the procedure effectively compensates for the different levels of occurrence of the categories. (More formally, the chance of the chi-squared distance for measuring relationships between profiles can be justified as a way of standardizing variables under a multinomial or Poisson distributional assumption; see Greenacre [1992].)

The complete set of chi-squared distances for all pairs of the five age groups, calculated as shown above, can be arranged in a matrix as follows:

$$\begin{array}{c} \\ 1 \\ 2 \\ 3 \\ 4 \\ 5 \end{array} \begin{array}{c} \begin{array}{ccccc} 1 & 2 & 3 & 4 & 5 \end{array} \\ \left(\begin{array}{ccccc} 0.00 & & & & \\ 0.09 & 0.00 & & & \\ 0.06 & 0.19 & 0.00 & & \\ 0.66 & 0.59 & 0.41 & 0.00 & \\ 1.07 & 1.01 & 0.83 & 0.51 & 0.00 \end{array} \right) \end{array}$$

The points representing the age groups in Display 16.5 give the two-dimensional representation of these distances, the *Euclidean distance*, between two points representing the chi-square distance between the corresponding age groups. (Similarly for the point representing type of relationship.) For a contingency table with r rows and c columns, it can be shown that the chi-squared distances can be represented *exactly* in $\min\{r - 1, c - 1\}$ dimensions; here, since $r = 3$ and $c = 5$, this means that the coordinates in Display 16.5 will lead to Euclidean distances that are *identical* to the chi-squared distances given above. For example, the correspondence analysis coordinates for age groups 1 and 2 taken from Display 16.4 are

Age Group	*x*	*y*
1	–0.355	–0.055
2	–0.290	0.000

The corresponding Euclidean distance is calculated as:

$$\sqrt{(-0.355 + 0.290)^2 + (-0.055 - 0.000)^2}$$

that is, a value of 0.09 — agreeing with the chi-squared distance between the two age groups given previously (Eq. (16.7).

Of most interest in correspondence analysis solutions such as that graphed in Display 16.5 is the *joint* interpretation of the points representing the row and column categories. It can be shown that row and column coordinates that are large and of the same sign correspond to a large positive residual term in the contingency table. Row and column coordinates that are large but of *opposite* signs imply a cell in the table with a large *negative* residual. Finally, small coordinate values close to the origin correspond to small residuals. In Display 16.5, for example, age group 5 and boyfriend/sexual intercourse both have large positive coordinate values on the first dimension. Consequently, the corresponding cell in the table will have a large positive residual. Again, age group 5 and boyfriend/no sexual intercourse have coordinate values with opposite signs on both dimensions, implying a negative residual for the corresponding cell in the table.

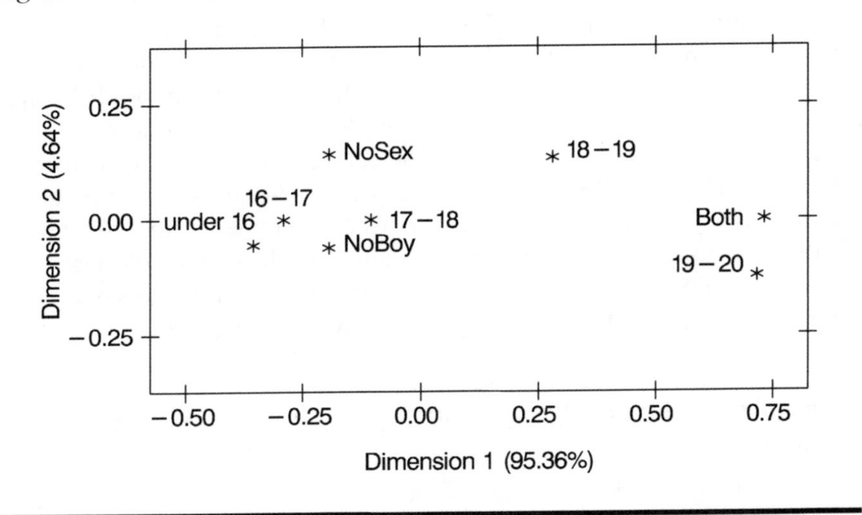

Display 16.5

16.3.2 Smoking and Motherhood

Assuming the cell counts of Display 16.2 are in an ASCII file births.dat and tab separated, they may be read in as follows:

```
data births;
   infile 'n:\handbook2\datasets\births.dat' expandtabs;
   input c1-c4;
   length rowid $12.;
   select(_n_);
      when(1) rowid='Young NS';
      when(2) rowid='Young Smoker';
```

```
      when(3) rowid='Old NS';
      when(4) rowid='Old Smoker';
   end;
   label c1='Prem Died' c2='Prem Alive' c3='FT Died'
   c4='FT Alive';
run;
```

As with the previous example, the data are read into a set of variables corresponding to the columns of the contingency table, and labels assigned to them. The character variable **rowid** is assigned appropriate values, using the automatic SAS variable **_n_** to label the rows. This is explicitly declared as a 12-character variable with the **length** statement. Where a character variable is assigned values as part of the data step, rather than reading them from a data file, the default length is determined from its first occurrence. In this example, that would have been from **rowid='Young NS'**; and its length would have been 8 with longer values truncated. This example also shows the use of the **select** group as an alternative to multiple **if-then** statements. The expression in parentheses in the **select** statement is compared to those in the **when** statements and the **rowid** variable set accordingly. The **end** statement terminates the select group.

The correspondence analysis and plot are produced in the same way as for the first example. The output is shown in Display 16.6 and the plot in Display 16.7.

```
proc corresp data=births out=coor;
   var c1-c4;
   id rowid;
run;

%plotit(data=coor,datatype=corresp,color=black,colors=black);
```

```
                        The CORRESP Procedure

               Inertia and Chi-Square Decomposition

Singular Principal     Chi-              Cumulative
   Value    Inertia  Square  Percent       Percent   18   36   54   72  90
                                                  ----+----+----+----+----+---
0.05032   0.00253  17.3467     90.78         90.78 ************************
0.01562   0.00024   1.6708      8.74         99.52 **
0.00365   0.00001   0.0914      0.48        100.00

   Total  0.00279  19.1090   100.00

Degrees of Freedom = 9
```

Row Coordinates

	Dim1	Dim2
Young NS	-0.0370	-0.0019
Young Smoker	0.0427	0.0523
Old NS	0.0703	-0.0079
Old Smoker	0.1042	-0.0316

Summary Statistics for the Row Points

	Quality	Mass	Inertia
Young NS	1.0000	0.6424	0.3158
Young Smoker	0.9988	0.0750	0.1226
Old NS	0.9984	0.2622	0.4708
Old Smoker	0.9574	0.0204	0.0908

Partial Contributions to Inertia for the Row Points

	Dim1	Dim2
Young NS	0.3470	0.0094
Young Smoker	0.0540	0.8402
Old NS	0.5114	0.0665
Old Smoker	0.0877	0.0839

Indices of the Coordinates that Contribute Most to Inertia for the Row Points

	Dim1	Dim2	Best
Young NS	1	0	1
Young Smoker	0	2	2
Old NS	1	0	1
Old Smoker	0	0	1

Squared Cosines for the Row Points

	Dim1	Dim2
Young NS	0.9974	0.0026
Young Smoker	0.3998	0.5990
Old NS	0.9860	0.0123
Old Smoker	0.8766	0.0808

The CORRESP Procedure

Column Coordinates

	Dim1	Dim2
Prem Died	0.3504	-0.0450
Prem Alive	0.0595	-0.0010
FT Died	0.2017	0.1800
FT Alive	-0.0123	-0.0005

Summary Statistics for the Column Points

	Quality	Mass	Inertia
Prem Died	0.9991	0.0152	0.6796
Prem Alive	0.9604	0.0749	0.0990
FT Died	0.9996	0.0066	0.1722
FT Alive	0.9959	0.9034	0.0492

Partial Contributions to Inertia for the Column Points

	Dim1	Dim2
Prem Died	0.7359	0.1258
Prem Alive	0.1047	0.0003
FT Died	0.1055	0.8730
FT Alive	0.0539	0.0008

Indices of the Coordinates that Contribute Most to Inertia for the Column Points

	Dim1	Dim2	Best
Prem Died	1	0	1
Prem Alive	0	0	1
FT Died	2	2	2
FT Alive	0	0	1

Squared Cosines for the Column Points

	Dim1	Dim2
Prem Died	0.9829	0.0162
Prem Alive	0.9601	0.0003
FT Died	0.5563	0.4433
FT Alive	0.9945	0.0014

Display 16.6

The chi-squared statistic for these data is 19.1090, which with nine degrees of freedom has an associated P-value of 0.024. Thus, it appears that "type" of mother is related to what happens to the newborn baby. The correspondence analysis of the data shows that the first two eigenvalues account for 99.5% of the inertia. Clearly, a two-dimensional solution provides an extremely good representation of the relationship between the two variables. The two-dimensional solution plotted in Display 16.7 suggests that young mothers who smoke tend to produce more full-term babies who then die in the first year, and older mothers who smoke have rather more than expected premature babies who die in the first year. It does appear that smoking is a risk factor for death in the first year of the baby's life and that age is associated with length of gestation, with older mothers delivering more premature babies.

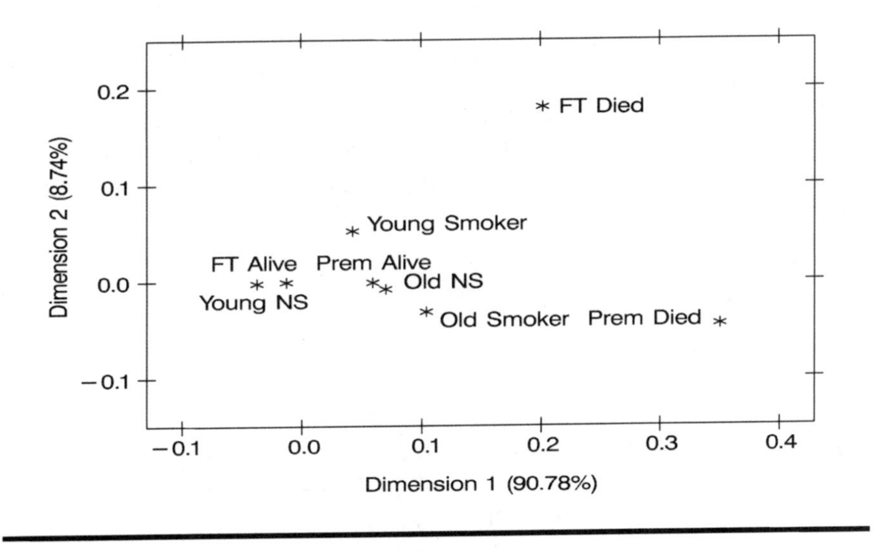

Display 16.7

16.3.3 Are the Germans Really Arrogant?

The data on perceived characteristics of European nationals can be read in as follows.

```
data europeans;
   infile 'n:\handbook2\datasets\europeans.dat' expandtabs;
   input country $ c1-c13;
   label c1='stylish'
```

```
        c2='arrogant'
        c3='sexy'
        c4='devious'
        c5='easy-going'
        c6='greedy'
        c7='cowardly'
        c8='boring'
        c9='efficient'
        c10='lazy'
        c11='hard working'
        c12='clever'
        c13='courageous';
run;
```

In this case, we assume that the name of the country is included in the data file so that it can be read in with the cell counts and used to label the rows of the table. The correspondence analysis and plot are produced in the same way and the results are shown in Displays 16.8 and 16.9.

```
proc corresp data=europeans out=coor;
   var c1-c13;
   id country;
run;

%plotit(data=coor,datatype=corresp,color=black,colors=black);
```

```
                          The CORRESP Procedure

                  Inertia and Chi-Square Decomposition

Singular  Principal    Chi-              Cumulative
   Value    Inertia  Square  Percent        Percent   10   20   30   40   50
                                                   ----+----+----+----+----+---
 0.49161    0.24168 255.697    49.73          49.73 *************************
 0.38474    0.14803 156.612    30.46          80.19 ***************
 0.20156    0.04063  42.985     8.36          88.55 ****
 0.19476    0.03793  40.133     7.81          96.36 ****
 0.11217    0.01258  13.313     2.59          98.95 *
 0.07154    0.00512   5.415     1.05         100.00 *

   Total    0.48597 514.153   100.00

Degrees of Freedom = 72
```

Row Coordinates

	Dim1	Dim2
France	0.6925	-0.3593
Spain	0.2468	0.3503
Italy	0.6651	0.0062
U.K.	-0.3036	0.1967
Ireland	-0.1963	0.6202
Holland	-0.5703	-0.0921
Germany	-0.5079	-0.5530

Summary Statistics for the Row Points

	Quality	Mass	Inertia
France	0.9514	0.1531	0.2016
Spain	0.4492	0.1172	0.0986
Italy	0.9023	0.1389	0.1402
U.K.	0.7108	0.1786	0.0677
Ireland	0.7704	0.1361	0.1538
Holland	0.5840	0.1002	0.1178
Germany	0.9255	0.1758	0.2204

Partial Contributions to Inertia for the Row Points

	Dim1	Dim2
France	0.3039	0.1335
Spain	0.0295	0.0971
Italy	0.2543	0.0000
U.K.	0.0681	0.0467
Ireland	0.0217	0.3537
Holland	0.1348	0.0057
Germany	0.1876	0.3632

Indices of the Coordinates that Contribute Most to Inertia for the Row Points

	Dim1	Dim2	Best
France	1	1	1
Spain	0	0	2
Italy	1	0	1
U.K.	0	0	1
Ireland	0	2	2
Holland	1	0	1
Germany	2	2	2

The CORRESP Procedure

Squared Cosines for the Row Points

	Dim1	Dim2
France	0.7496	0.2018
Spain	0.1490	0.3002
Italy	0.9023	0.0001
U.K.	0.5007	0.2101
Ireland	0.0701	0.7002
Holland	0.5692	0.0148
Germany	0.4235	0.5021

Column Coordinates

	Dim1	Dim2
stylish	0.8638	-0.3057
arrogant	-0.0121	-0.5129
sexy	0.9479	-0.1836
devious	0.2703	0.0236
easy-going	0.0420	0.5290
greedy	0.1369	-0.1059
cowardly	0.5869	0.2627
boring	-0.2263	0.0184
efficient	-0.6503	-0.4192
lazy	0.2974	0.5603
hard working	-0.4989	-0.0320
clever	-0.5307	-0.2961
courageous	-0.4976	0.6278

Summary Statistics for the Column Points

	Quality	Mass	Inertia
stylish	0.9090	0.0879	0.1671
arrogant	0.6594	0.1210	0.0994
sexy	0.9645	0.0529	0.1053
devious	0.3267	0.0699	0.0324
easy-going	0.9225	0.1248	0.0784
greedy	0.2524	0.0473	0.0115
cowardly	0.6009	0.0350	0.0495
boring	0.4431	0.0633	0.0152
efficient	0.9442	0.1040	0.1356
lazy	0.6219	0.0671	0.0894
hard working	0.9125	0.1361	0.0767
clever	0.9647	0.0227	0.0179
courageous	0.7382	0.0681	0.1217

Partial Contributions to Inertia for the Column Points

	Dim1	Dim2
stylish	0.2714	0.0555
arrogant	0.0001	0.2150
sexy	0.1968	0.0121
devious	0.0211	0.0003
easy-going	0.0009	0.2358
greedy	0.0037	0.0036
cowardly	0.0499	0.0163
boring	0.0134	0.0001
efficient	0.1819	0.1234
lazy	0.0246	0.1423
hard working	0.1401	0.0009
clever	0.0264	0.0134
courageous	0.0697	0.1812

The CORRESP Procedure

Indices of the Coordinates that Contribute Most to Inertia for the Column Points

	Dim1	Dim2	Best
stylish	1	0	1
arrogant	0	2	2
sexy	1	0	1
devious	0	0	1
easy-going	0	2	2
greedy	0	0	1
cowardly	0	0	1
boring	0	0	1
efficient	1	1	1
lazy	0	2	2
hard working	1	0	1
clever	0	0	1
courageous	2	2	2

Squared Cosines for the Column Points

	Dim1	Dim2
stylish	0.8078	0.1012
arrogant	0.0004	0.6591
sexy	0.9296	0.0349
devious	0.3242	0.0025
easy-going	0.0058	0.9167
greedy	0.1579	0.0946
cowardly	0.5006	0.1003
boring	0.4402	0.0029
efficient	0.6670	0.2772
lazy	0.1367	0.4851
hard working	0.9088	0.0038
clever	0.7357	0.2290
courageous	0.2848	0.4534

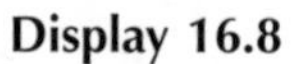

Display 16.8

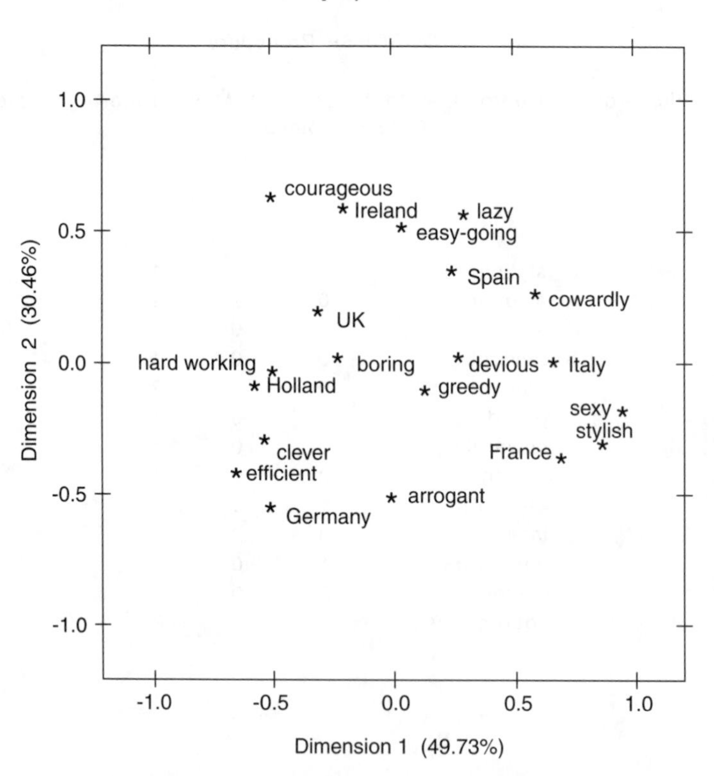

Display 16.9

Here, a two-dimensional representation accounts for approximately 80% of the inertia. The two-dimensional solution plotted in Display 16.9 is left to the reader for detailed interpretation, noting only that it largely fits the author's own prejudices about perceived national stereotypes.

Exercises

16.1 Construct a scatterplot matrix of the first four correspondence analysis coordinates of the European stereotypes data.

16.2 Calculate the chi-squared distances for both the row and column profiles of the smoking and motherhood data, and then compare them with the corresponding Euclidean distances in Display 16.7.

Appendix A

SAS Macro to Produce Scatterplot Matrices

This macro is based on one supplied with the SAS system but has been adapted and simplified. It uses proc iml and therefore requires that SAS/IML be licensed.

The macro has two arguments: the first is the name of the data set that contains the data to be plotted; the second is a list of numeric variables to be plotted. Both arguments are required.

```
%macro scattmat(data,vars);

/* expand variable list and separate with commas */

data _null_;
   set &data (keep=&vars);
   length varlist $500. name $32.;
   array xxx {*} _numeric_;
   do i=1 to dim(xxx);
      call vname(xxx{i},name);
      varlist=compress(varlist||name);
      if i<dim(xxx) then varlist=compress(varlist||',');
   end;
   call symput('varlist',varlist);
```

```
   stop;
run;

proc iml;
      /*-- Load graphics --*/
      call gstart;

      /*-- Module : individual scatter plot --*/
      start gscatter(t1, t2);
         /* pairwise elimination of missing values */
         t3=t1;
         t4=t2;
         t5=t1+t2;
         dim=nrow(t1);
         j=0;
         do i=1 to dim;
            if t5[i]=. then ;
            else do;
            j=j+1;
            t3[j]=t1[i];
            t4[j]=t2[i];
            end;
         end;
         t1=t3[1:j];
         t2=t4[1:j];
         /* ---------------- */
         window=(min(t1)||min(t2))//
                  (max(t1)||max(t2));
         call gwindow(window);
         call gpoint(t1,t2);
      finish gscatter;

   /*-- Module : do scatter plot matrix --*/
   start gscatmat(data, vname);
         call gopen('scatter');
         nv=ncol(vname);
         if (nv=1) then nv=nrow(vname);
         cellwid=int(90/nv);
         dist=0.1*cellwid;
```

```
      width=cellwid-2*dist;
      xstart=int((90 -cellwid * nv)/2) + 5;
      xgrid=((0:nv-1)#cellwid + xstart)`;

/*-- Delineate cells --*/
      cell1=xgrid;
      cell1=cell1||(cell1[nv]//cell1[nv-(0:nv-2)]);
      cell2=j(nv, 1, xstart);
      cell2=cell1[,1]||cell2;
      call gdrawl(cell1, cell2);
      call gdrawl(cell1[,{2 1}], cell2[,{2 1}]);
      xstart = xstart + dist;  ystart = xgrid[nv] + dist;

/*-- Label variables ---*/
      call gset("height", 3);
      call gset("font","swiss");
      call gstrlen(len, vname);
      where=xgrid[1:nv] + (cellwid-len)/2;
      call gscript(where, 0, vname) ;
      len=len[nv-(0:nv-1)];
      where=xgrid[1:nv] + (cellwid-len)/2;
      call gscript(0,where, vname[nv - (0:nv-1)]);

/*-- First viewport --*/
      vp=(xstart || ystart)//((xstart || ystart) + width) ;

/*   Since the characters are scaled to the viewport   */
/*   (which is inversely porportional to the           */
/*   number of variables),                             */
/*   enlarge it proportional to the number of variables */

   ht=2*nv;
   call gset("height", ht);
   do i=1 to nv;
      do j=1 to i;
          call gportstk(vp);
        if (i=j) then ;
           else run gscatter(data[,j], data[,i]);

      /*-- onto the next viewport --*/
```

```
            vp[,1] = vp[,1] + cellwid;
            call gportpop;
         end;
         vp=(xstart // xstart + width) || (vp[,2] - cellwid);
      end;
      call gshow;
  finish gscatmat;

     /*-- Placement of text is based on the character
height.              */
     /* The IML modules defined here assume percent as the
unit of   */
     /* character height for device independent
control.                    */
  goptions gunit=pct;

  use &data;
  vname={&varlist};
  read all var vname into xyz;
  run gscatmat(xyz, vname);
  quit;

  goptions gunit=cell;          /*-- reset back to default --*/
%mend;
```

Appendix B

Answers to Selected Chapter Exercises

The answers given here assume that the data have been read in as described in the relevant chapter.

Chapter 2

```
proc univariate data=water normal plot;                /* 2.1 */
   var mortal hardness;
run;

proc sort data=water;                                  /* 2.2 */
   by location;
run;
proc boxplot data=water;
   plot (mortal hardness)*location;
run;

proc univariate data=water normal;                     /* 2.3 */
   var hardness;
   histogram hardness / lognormal(theta=est) exponential
(theta=est);
```

```
    probplot hardness / lognormal(theta=est sigma=est
zeta=est);
    probplot hardness / exponential (theta=est sigma=est);
  run;

  proc univariate data=water;                          /* 2.4 */
    var mortal hardness;
    histogram mortal hardness /kernel ;
  run;

  proc sort data=water;                                /* 2.5 */
    by location;
  run;
  proc kde data=water out=bivest;
    var mortal hardness;
    by location;
  run;
  proc g3d data=bivest;
    plot hardness*mortal=density;
    by location;
  run;

  proc gplot data=water;                               /* 2.6 */
    plot mortal*hardness=location;
    symbol1 v=dot i=r l=1;
    symbol2 v=circle i=r l=2;
  run;
```

Chapter 3

```
  data pill2;                                          /* 3.1 */
    set the_pill;
    use=caseuse;
    case='Y';
    output;
    use=contruse;
    case='N';
    output;
  run;
```

```
proc freq data=pill2;
   tables case*use /riskdiff;
   weight n;
run;
```

The short data step restructures the data into separate observations for cases and controls rather than case-control pairs enumerated in Display 3.4.

```
proc freq data=pistons order=data;                    /* 3.2 */
   tables machine*site / out=tabout outexpect outpct;
   weight n;
run;
data resids;
   set tabout;
   r=(count-expected)/sqrt(expected);
   radj=r/sqrt((1-percent/pct_row)*(1-percent/pct_col));
run;
proc tabulate data=resids;
   class machine site;
   var r radj;
   table machine,
      site*r;
   table machine,
      site*radj;
run;

data lesions2;                                        /* 3.3 */
   set lesions;
   region2=region;
   if region ne 'Gujarat' then region2='Others';
run;
proc freq data=lesions2 order=data;
   tables site*region2 /exact;
   weight n;
   run;
```

Chapter 4

```
proc reg data=uscrime;                                /* 4.1 */
```

```
   model R= Age--Ed Ex1--X / selection=cp;
run;

proc reg data=uscrime;                                   /* 4.2 */
   model R= Age Ed Ex1 U2 X / selection=cp start=1 stop=5;
   plot cp.*np. / cmallows=black;
run;
```

When selection=cp is used, the start and stop options determine the smallest and largest number of explanatory variables to be included in the model.

```
data uscrime;                                            /* 4.5 */
   set uscrime;
   age_s=age*s;
run;

proc reg data=uscrime;
   model R=age s age_s;
   output out=regout p=rhat;
run;

proc gplot data=regout;
   plot rhat*age=s / vaxis=axis1;
   plot2 r*age=s /vaxis=axis1;
   symbol1 i=join v=none l=1;
   symbol2 i=join v=none l=2;
   symbol3 v=dot;
   symbol4 v=circle;
   axis1 order=20 to 200 by 10;
run;
```

Because proc reg has no facility for specifying interactions in the model statement, the short data step computes a new variable age_s to be used as an interaction term.

The plot2 statement is used as a "trick" to overlay two y*x=z type plots. To ensure that the vertical axes are kept in alignment the plot and plot2 statements, both use the same vertical axis definition.

Chapter 5

```
proc anova data=hyper;                                   /* 5.1 */
   class diet drug biofeed;
   model logbp=diet drug biofeed;
   means drug / bon duncan;
run;

proc sort data=hyper;                                    /* 5.2 */
   by diet;
proc boxplot data=hyper;
   plot logbp*diet;
run;
proc sort data=hyper;
   by drug;
proc boxplot data=hyper;
   plot logbp*drug;
run;
proc sort data=hyper;
   by biofeed;
proc boxplot data=hyper;
   plot logbp*biofeed;
run;
```

Chapter 6

6.1

Try using @ to restrict the expansion of the bar operator; for example,

```
model days=type|sex|origin|grade@3;
```

or

```
model days=type|sex|origin|grade@2;
```

and then specifying additional terms.

```
data ozkids;                                   /* 6.2 */
   set ozkids;
   logdays=log(days+0.5);
run;
```

Add 0.5 to allow for zeros.

```
proc tabulate data=ozkids f=6.2;               /* 6.3 */
   class type sex origin grade;
   var days;
   table sex*grade,
      origin*type*days*(mean std);
run;

proc glm data=ozkids noprint;                  /* 6.4 */
   class origin sex grade type;
   model days=origin sex grade type;
   output out=glmout r=res;
run;

proc univariate data=glmout noprint;
   var res;
   probplot;
run;
```

Chapter 7

For exercises 7.1 and 7.2, the data need to be restructured so that each measurement is an observation.

```
data vision2;
     set vision;
     array xall {8} x1-x8;
     do i=1 to 8;
        if i > 4 then eye='R';
              else eye='L';
        select(i);
           when(1,5) strength=1;
```

```
        when(2,6) strength=3;
        when(3,7) strength=6;
        when(4,8) strength=10;
      end;
    response=xall{i};
    output;
    end;
    drop i x1-x8;
run;

proc gplot data=vision2;                          /* 7.1 */
    plot response*strength=eye;
    symbol1 i=std1mj l=1;
    symbol2 i=std1mj l=2;
    run;

proc sort data=vision2;                           /* 7.2 */
    by eye strength;
run;

proc boxplot data=vision2;
    plot response*strength;
    by eye;
run;

proc corr data=vision;                            /* 7.3 */
    var x1-x8;
run;
```

Chapter 8

```
proc logistic data=ghq;                           /* 8.1 */
    class sex;
    model cases/total=ghq sex ghq*sex;
run;

proc logistic data=plasma desc;                   /* 8.2 */
    model esr=fibrinogen|fibrinogen gamma|gamma;
run;
```

Chapter 9

```
proc genmod data=ozkids;                                   /* 9.1 */
   class origin sex grade type;
   model days=sex origin type grade grade*origin / dist=p
link=log type1 type3 scale=3.1892;
run;

data ozkids;                                               /* 9.2 */
   set ozkids;
   absent=days>13;
run;

proc genmod data=ozkids desc;
   class origin sex grade type;
   model absent=sex origin type grade grade*origin / dist=b
link=logit type1 type3;
run;

proc genmod data=ozkids desc;
   class origin sex grade type;
   model absent=sex origin type grade grade*origin / dist=b
link=probit type1 type3;
run;
```

Chapter 10

```
data pndep2;                                               /* 10.1 */
   set pndep2;
   depz=dep;
run;

proc sort data=pndep2;
   by idno time;
run;

proc stdize data=pndep2 out=pndep2;
   var depz;
   by idno;
run;
```

First, a new variable is created to hold the standardized depression scores. Then the standardization is done separately for each subject.

```
goptions reset=symbol;
symbol1 i=join v=none l=1 r=27;
symbol2 i=join v=none l=2 r=34;
proc gplot data=pndep2;
   plot depz*time=idno /nolegend skipmiss;
run;

data pndep2;                                        /* 10.2 */
   set pndep2;
   if time=1 then time=2;
run;
```

The two baseline measures are treated as if they were made at the same time.

```
proc sort data=pndep2;
   by idno;
run;

proc reg data=pndep2 outest=regout(keep=idno time)
noprint;
   model dep=time;
   by idno;
run;
```

A separate regression is run for each subject and the slope estimate saved. This is renamed as it is merged into the **pndep** data set so that the variable **time** is not overwritten.

```
data pndep;
   merge pndep regout (rename=(time=slope));
   by idno;
run;

proc ttest data=pndep;
   class group;
   var slope;
run;
```

```
proc glm data=pndep;
   class group;
   model slope=mnbase group /solution;
run;
```

Chapter 11

```
proc mixed data=alzheim method=ml covtest;          /* 11.1 */
   class group idno;
   model score=group visit group*visit /s ;
   random int /subject=idno type=un;
run;

proc mixed data=alzheim method=ml covtest;
   class group idno;
   model score=group visit group*visit /s ;
   random int visit /subject=idno type=un;
run;

proc sort data=alzheim;                              /* 11.2 */
   by idno;
run;

proc reg data=alzheim outest=regout(keep=idno intercept
visit) noprint  ;
   model score=visit;
   by idno;
run;

data regout;
   merge regout(rename=(visit=slope)) alzheim;
   by idno;
   if first.idno;
run;

proc gplot data=regout;
   plot intercept*slope=group;
   symbol1 v='L';
   symbol2 v='P';
run;
```

```
data pndep(keep=idno group x1-x8)
      pndep2(keep=idno group time dep mnbase);  /* 11.3 */
   infile 'n:\handbook2\datasets\channi.dat';
   input group x1-x8;
   idno=_n_;
   mnbase=mean(x1,x2);
   if x1=-9 or x2=-9 then mnbase=max(x1,x2);
   array xarr {8} x1-x8;
do i=1 to 8;
   if xarr{i}=-9 then xarr{i}=.;
      time=i;
      dep=xarr{i};
      output pndep2;
   end;
   output pndep;
run;
```

The data step is rerun to include the mean of the baseline measures in pndep2, the data set with one observation per measurement. A where statement is then used with the proc step to exclude the baseline observations from the analysis.

```
proc mixed data=pndep2 method=ml covtest;
   class group idno;
   model dep=mnbase time group /s;
   random int /sub=idno;
   where time>2;
run;

proc mixed data=pndep2 method=ml covtest;
   class group idno;
   model dep=mnbase time group /s;
   random int time /sub=idno type=un;
   where time>2;
run;

data alzheim;                                    /* 11.4 */
   set alzheim;
   mnscore=mean(of score1-score5);
   maxscore=max(of score1-score5);
```

```
run;

proc ttest data=alzheim;
   class group;
      var mnscore maxscore;
      where visit=1;
run;
```

Chapter 12

Like proc reg, proc phreg has no facility for specifying categorical predictors or interaction terms on the model statement. Additional variables must be created to represent these terms. The following steps censor times over 450 days and create suitable variables for the exercises.

```
data heroin3;
   set heroin;
   if time > 450 then do;                    /* censor times over 450 */
      time=450;
      status=0;
      end;
   clinic=clinic-1;                           /* recode clinic to 0,1  */
   dosegrp=1;                             /* recode dose to 3 groups */
   if dose >= 60 then dosegrp=2;
   if dose >=80 then dosegrp=3;
   dose1=dosegrp eq 1;              /* dummies for dose group */
   dose2=dosegrp eq 2;
   clindose1=clinic*dose1;            /* dummies for interaction */
   clindose2=clinic*dose2;
run;
proc stdize data=heroin3 out=heroin3;
   var dose;
run;
data heroin3;
   set heroin3;
   clindose=clinic*dose;                          /* interaction term */
run;

proc phreg data=heroin3;                                     /* 12.1 */
   model time*status(0)=prison dose clinic / rl;
```

```
run;

data covvals;                                                    /* 12.2 */
   retain prison clinic 0;
   input dose1 dose2;
cards;
1 0
0 1
0 0
;
proc phreg data=heroin3;
   model time*status(0)=prison clinic dose1 dose2 / rl;
   baseline covariates=covvals out=baseout  survival=bs /
method=ch nomean;
run;
data baseout;
   set baseout;
   dosegrp=3-dose1*2-dose2;
run;
proc gplot data=baseout;
   plot bs*time=dosegrp;
   symbol1 v=none i=stepjs l=1;
   symbol2 v=none i=stepjs l=3;
   symbol3 v=none i=stepjs l=33 w=20;
run;
```

By default, the **baseline** statement produces survival function estimates at the mean values of the covariates. To obtain estimates at specific values of the covariates, a data set is created where the covariates have these values, and it is named in the **covariates=** option of the **baseline** statement. The covariate values data set must have a corresponding variable for each predictor in the **phreg** model.

The survival estimates are plotted as step functions using the **step** interpolation in the **symbol** statements. The j suffix specifies that the steps are to be joined and the **s** suffix sorts the data by the x-axis variable.

```
proc phreg data=heroin3;                                         /* 12.3 */
   model time*status(0)=prison clinic dose1 dose2 clindose1
clindose2/ rl;
   test clindose1=0, clindose2=0;
run;
```

```
proc phreg data=heroin3;
   model time*status(0)=prison clinic dose clindose / rl;
run;

proc phreg data=heroin3;                                    /* 12.4 */
   model time*status(0)=prison clinic dose / rl;
   output out=phout xbeta=lp resmart=resm resdev=resd;
run;

goptions reset=symbol;
proc gplot data=phout;
   plot (resm resd)*lp / vref=-2,2 lvref=2;
   symbol1 v=circle;
run;
```

Chapter 13

```
data decathlon;                                             /* 13.1 */
   infile 'n:handbook2\datasets\olympic.dat' expandtabs;
   input run100 Ljump shot Hjump run400 hurdle discus
polevlt javelin run1500 score;
   run100=run100*-1;
   run400=run400*-1;
   hurdle=hurdle*-1;
   run1500=run1500*-1;
run;

/* and re-run analysis as before */

proc princomp data=pain;                                    /* 13.2 */
   var p1-p9;
run;

proc factor data=decathlon method=principal priors=smc
mineigen=1 rotate=oblimin;                                  /* 13.3 */
   var run100--run1500;
   where score>6000;
run;
```

```
   proc factor data=decathlon method=ml min=1 rotate=obvari
max;
      var run100--run1500;
      where score>6000;
   run;
```

Chapter 14

```
   proc modeclus data=usair2 out=modeout method=1 std
r=1 to 3 by .25  test;                                  /* 14.1 */
      var temperature--rainydays;
      id city;
   run;
   proc print data=modeout;
      where _R_=2;
   run;
```

The **proc** statement specifies a range of values for the kernel radius based on the value suggested as a "reasonable first guess" in the documentation. The **test** option is included for illustration, although it should not be relied on with the small numbers in this example.

```
   proc stdize data=usair2 method=range out=usair3; /* 14.3 */
      var temperature--rainydays;
   run;

   /* then repeat the analyses without the std option */

   proc glm data=clusters;                              /* 14.4 */
      class cluster;
      model so2=cluster;
      means cluster / scheffe;
   run;
```

Chapter 15

15.2 The first example uses a nearest neighbour method with k = 4 and the second a kernel method with a value of the smoothing parameter derived from a formula due to Epanechnikov (Epanechnikov [1969]).

```
proc discrim data=skulls method=npar k=4 crossvalidate;
                                                  /* 15.2 */
   class type;
   var length--facewidth;
run;

proc discrim data=skulls method=npar kernel=normal r=.7
crossvalidate;
   class type;
   var length--facewidth;
run;
```

Chapter 16

16.1 Use the scatterplot matrix macro introduced in Chapter 4.

```
proc corresp data=europeans out=coor dim=4;      /* 16.1 */
   var c1-c13;
   id country;
run;

%include 'scattmat.sas';
%scattmat(coor,dim1-dim4);
```

16.2 The cp and rp options (row profile and column profile), together with the row and column masses, provide the proportions used in the calculations.

```
proc corresp data=births out=coor cp rp;         /* 16.2 */
   var c1-c4;
   id rowid;
run;
```

References

Agresti, A. (1996) *Introduction to Categorical Data Analysis*, Wiley, New York.

Aitkin, M. (1978) The analysis of unbalanced cross-classifications. *Journal of the Royal Statistical Society* A, 141, 195–223.

Akaike, H. (1974) A new look at the statistical model identification. *MEE Transactions in Automatic Control*, 19, 716–723.

Berk, K.N. (1977) Tolerance and condition in regression computations. *Journal of the American Statistical Association*, 72, 863–866.

Box, G.E.P. and Cox, D.R. (1964) An analysis of transformations (with discussion). *Journal of the Royal Statistical Society A*, 143, 383–430.

Caplehorn, J. and Bell, J. (1991) Methadone dosage and the retention of patients in maintenance treatment. *The Medical Journal of Australia*, 154, 195–199.

Chatterjee, S. and Prize, B. (1991) *Regression Analysis by Example* (2nd edition), Wiley, New York.

Clayton, D. and Hills, M. (1993) *Statistical Models in Epidemiology*, Oxford University Press, Oxford.

Collett, D. (1994) *Modelling Survival Data in Medical Research*, CRC/Chapman & Hall, London.

Collett, D. (1991) *Modelling Binary Data*, CRC/Chapman & Hall, London.

Cook, R.D. (1977) Detection of influential observations in linear regression. *Technometrics*, 19, 15–18.

Cook, R.D. (1979) Influential observations in linear regression. *Journal of the American Statistical Association*, 74, 169–174.

Cook, R.D. and Weisberg, S. (1982) *Residuals and Influence in Regression*, CRC/Chapman & Hall, London.

Cox, D.R. (1972) Regression models and life tables. *Journal of the Royal Statistical Society* B, 34, 187–220.

Crowder, M.J. and Hand, D.J. (1990) *Analysis of Repeated Measures*, CRC/Chapman & Hall, London.

Davidson, M.L. (1972) Univariate versus multivariate tests in repeated measurements experiments. *Psychological Bulletin*, 77, 446–452.

Diggle, P.L., Liang, K., and Zeger, S.L. (1994) *Analysis of Longitudinal Data,* Oxford University Press, Oxford.

Dixon, W.J. and Massey, F.J. (1983) *Introduction to Statistical Analysis*, McGraw-Hill, New York.

Dizney, H. and Groman, L. (1967) Predictive validity and differential achievement in three MLA comparative foreign language tests. *Educational and Psychological Measurement*, 27, 1127–1130.

Epanechnikov, V.A. (1969) Nonparametric estimation of a multivariate probability density. *Theory of Probability and its Applications*, 14, 153–158.

Everitt, B.S. (1987) *An Introduction to Optimization Methods and their Application in Statistics*, CRC/Chapman & Hall, London.

Everitt, B.S. (1992) *The Analysis of Contingency Tables* (2nd edition), CRC/Chapman & Hall, London.

Everitt, B.S. (1998) *The Cambridge Dictionary of Statistics*, Cambridge University Press, Cambridge.

Everitt, B.S. (2001) *Statistics in Psychology: An Intermediate Course.* Laurence Erlbaum, Mahwah, New Jersey.

Everitt, B.S. and Pickles, A. (2000) *Statistical Aspects of the Design and Analysis of Clinical Trials*, ICP, London.

Everitt, B.S. and Dunn, G. (2001) *Applied Multivariate Data Analysis* (2nd edition), Edward Arnold, London.

Everitt, B.S., Landau, S., and Leese, M. (2001) *Cluster Analysis* (4th edition), Edward Arnold, London.

Fisher, R.A. (1936) The use of multiple measurement in taxonomic problems. *Annals of Eugenics*, 7, 179–184.

Fisher, L.D. and van Belle, G. (1993) *Biostatistics: A Methodology for the Health Sciences*, Wiley, New York.

Goldberg, D. (1972) *The Detection of Psychiatric Illness by Questionnaire*, Oxford University Press, Oxford.

Greenacre, M. (1984) *Theory and Applications of Correspondence Analysis*, Academic Press, Florida.

Greenacre, M. (1992) Correspondence analysis in medical research. *Statistical Methods in Medical Research*, 1, 97–117.

Greenhouse, S.W. and Geisser, S. (1959) On methods in the analysis of profile data. *Psychometrika*, 24, 95–112.

Gregoire, A.J.P., Kumar, R., Everitt, B.S., Henderson, A.F., and Studd, J.W.W. (1996) Transdermal oestrogen for the treatment of severe post-natal depression. *The Lancet*, 347, 930–934.

Hand, D.J. (1981) *Discrimination and Classification*, Wiley, Chichester.

Hand, D.J. (1986) Recent advances in error rate estimation. *Pattern Recognition Letters*, 4, 335–346.

Hand, D.J. (1997) *Construction and Assessment of Classification Rules*, Wiley, Chichester.

Hand, D.J., Daly, F., Lunn, A.D., McConway, K.J., and Ostrowski, E. (1994) *A Handbook of Small Data Sets*, CRC/Chapman & Hall, London.

Hosmer, D.W. and Lemeshow, S. (1999) *Applied Survival Analysis*, Wiley, New York.

Hotelling, H. (1933) Analysis of a complex of statistical variables into principal components. *Journal of Educational Psychology*, 24, 417–441.

Howell, D.C. (1992) *Statistical Methods for Psychologists*, Duxbury Press, Belmont, California.

Huynh, H. and Feldt, L.S. (1976) Estimates of the Box correction for degrees of freedom for sample data in randomised block and split plot designs. *Journal of Educational Statistics*, 1, 69–82.

Kalbfleisch, J.D. and Prentice, J.L. (1980) *The Statistical Analysis of Failure Time Data*, Wiley, New York.

Krzanowski, W.J. and Marriott, F.H.C. (1995) *Multivariate Analysis*, Part 2, Edward Arnold, London.

Lawless, J.F. (1982) *Statistical Models and Methods for Lifetime Data*, Wiley, New York.

Levene, H. (1960) Robust tests for the equality of variance. *Contribution to Probability and Statistics* (O. Olkin, Ed.), Stanford University Press, California.

McCullagh, P. and Nelder, J.A. (1989) *Generalized Linear Models*, CRC/Chapman & Hall, London.

McKay, R.J. and Campbell, N.A. (1982a) Variable selection techniques in discriminant analysis. I. Description. *British Journal of Mathematical and Statistical Psychology*, 35, 1–29.

McKay, R.J. and Campbell, N.A. (1982b) Variable selection techniques in discriminant analysis. II. Allocation. *British Journal of Mathematical and Statistical Psychology*, 35, 30–41.

Mallows, C.L. (1973) Some comments on Cp. *Technometrics*, 15, 661–675.

Matthews, J.N.S., Altman, D.G., Campbell, M.J., and Royston, P. (1990) Analysis of serial measurements in medical research. *British Medical Journal*, 300, 230–235.

Matthews, J.N.S. (1993) A refinement to the analysis of serial data using summary measures. *Statistics in Medicine*, 12, 27–37.

Maxwell, S.E. and Delaney, H.D. (1990) *Designing Experiments and Analysing Data*, Wadsworth, Belmont, California.

Milligan, G.W. and Cooper, M.C. (1988) A study of standardization of variables in cluster analysis. *Journal of Classification*, 5, 181–204.

Nelder, J.A. (1977) A reformulation of linear models. *Journal of the Royal Statistical Society A*, 140, 48–63.

Pearson, K. (1901) On lines and planes of closest fit to systems of points in space. *Philosophical Magazine*, 2, 559–572.

Pinheiro, J.C. and Bates, D.M (2000) *Mixed-Effects Models in S and S-PLUS*, Springer, New York.

Quine, S. (1975) *Achievement Orientation of Aboriginal and White Adolescents.* Doctoral dissertation, Australian National University, Canberra.

Rouanet, H. and Lepine, D. (1970) Comparison between treatments in a repeated measures design: ANOVA and multivariate methods. *British Journal of Mathematical and Statistical Psychology*, 23, 147–163.

Sarle, W.S. (1983) SAS Technical Report A-108 Cubic Clustering Criterion, SAS Institute Inc., Cary, NC.

Sartwell, P.E., Mazi, A.T. Aertles, F.G., Greene, G.R., and Smith, M.E. (1969) Thromboembolism and oral contraceptives: an epidemiological case-control study. *American Journal of Epidemiology*, 90, 365–375.

Satterthwaite, F.W. (1946) An approximate distribution of estimates of variance components. *Biometrics Bulletin*, 2, 110–114.

Scheffé, H. (1959) *The Analysis of Variance*, Wiley, New York.

Senn, S. (1997) *Statistical Issues in Drug Development*, Wiley, Chichester.

Shapiro, S.S. and Wilk, M.B. (1965) An analysis of variance test for normality. *Biometrika*, 52, 591–611.

Silverman, B.W. (1986) *Density Estimation in Statistics and Data Analysis*, CRC/Chapman & Hall, London.

Skevington, S.M. (1990) A standardised scale to measure beliefs about controlling pain (B.P.C.Q.); a preliminary study. *Psychological Health*, 4, 221–232.

Somes, G.W. and O'Brien, K.F. (1985) Mantel-Haenszel statistic. *Encyclopedia of Statistical Sources*, Vol. 5 (S. Kotz, N.L. Johnson, and C.B. Read, Eds.), Wiley, New York.

Vandaele, W. (1978) Participation in illegitimate activities: Erlich revisited. *Deterrence and Incapacitation* (Blumstein, A., Cohen, J., and Nagin, D., Eds.), Natural Academy of Sciences, Washington, D.C.

Index

A

B

C

F

G

H

I

N

O

Q

R

S

T

U

V

W

Y